Forest H. Belt's

Easi-Guide
to
CONSERVING ENERGY
&
MATERIALS

Text by
Forest H. Belt, Marti McPherson,
and Pat Constable

Photography by
Forest H. Belt and Marti McPherson

HOWARD W. SAMS & CO., INC.
THE BOBBS-MERRILL CO., INC.
INDIANAPOLIS · KANSAS CITY · NEW YORK

Introduction

What to do about the energy "crisis" worries people all over the world these days. In the United States in particular, the idea of not having enough of anything leaves just about everyone uneasy. Most Americans are unaccustomed to thinking that way. They have not experienced serious shortages since World War 2.

If the situation hasn't affected you already, it will soon. The question is, what can you do about it? *There is plenty you can do.* This whole book devotes itself to showing and telling you what, in specific terms, and how. You'll find a wealth of ideas and tips you can use every day to improve your mode of living, even if the shortages don't yet cause you any major inconvenience.

About the "crisis." Some cynics believe it's politically contrived. Others insist it's a scheme of big business to raise profits. Both notions could be right, or both wrong. But knowing, and placing blame, does not in itself help much when you can't buy enough gasoline to get home from a trip. Or when your house is cold. Or when meat costs too much and chews like leather. Or when you can't find soap, or toilet paper, or nylon hose. In situations like those, your concern drifts from "who did it?" and turns more to "what do we do now?"

Those of us who put this book together for you have three broad suggestions. (1) Quit calling the situation a "crisis." What we have in the U.S. is merely a shortage situation (rather, several of them). There's no real crisis yet. (2) Having thus shaped your outlook, begin immediately coping with the shortages and, where you can, help alleviate them. That way, no situation will become a crisis. (3) Enlist friends and neighbors. Alert everyone you know to the sensible ways to make life easier in spite of shortages. Employ these three fundamental ideas, and you'll make a giant contribution to putting things back on even keel.

Many people helped make this book complete in its coverage of the various shortages. My two coauthors deserve a large share of credit. Both are professional women, mothers, and homemakers. Marti McPherson is a television writer/producer/hostess, known best for her own series "The Dwelling Place" and as former producer of "Indy Today," a popular local talk show. Pat Constable operates her own small ad agency, keeps a busy schedule at a Midwest radio station, and still finds time for plain and fancy cooking at home. Along with you and me, they deal every day with the country's shortages.

Several companies and organizations gave their interest and assistance. We thank in particular the management and employees of Amtrak, Family Camping Federation, General Mills Co., Grumman Allied Industries, Inc., The Hoover Company, Indiana Division of Tourism, Key Market (Jeffersonville IN), Libby-Owens-Ford, Marathon Oil Co., Pennsylvania House, Recreational Vehicle Institute, Skelly Oil Co., Spivey Construction Co., Stewart-Carey Inc., Trans World Airlines, and the University of Hawaii.

We received extraordinary help from people at Public Service Indiana, Hoosier Oil Co. (Indianapolis), Citizens Gas & Coke Utility, Dan Young Chevrolet (Indianapolis), Mobil Oil Company, and the Indiana Dept. of Conservation. They earned our—and your—special gratitude.

Where the energy situation goes from here depends on you as much as on anyone. You are the one who decides what you will buy, how much of it, who from, and—believe it or not—the price you will pay. Shortages of anything occur only when supply and demand don't coincide. Supplies may be inadequate or demand may be too heavy. Either one makes prices rise. The answer to shortages, therefore, does not lie in panic buying or selfish hoarding.

You'll find the practical answer to shortages in learning to conserve. As a concept, conservation is not new. As a practical matter, all of us could use help to see how we can make the most of what there is to buy. I and my coauthors anticipate that you will find solutions for your own personal shortage situations somewhere in these pages.

Forest H. Belt

Contents

Chapter 1

Ways to Have Enough Gasoline

The shortage of petroleum has spawned many myths about gasoline. One is that a small engine always burns less gasoline per mile. That's not true unless the car weighs less too.

A gallon of gasoline contains exactly so much energy. You push the accelerator and your car moves a certain number of miles before that gallon is used up. The weight of the car, no matter what the engine size, takes that much energy (horsepower) to move that far. In fact, if an engine is too small for the car's weight, you'd get lousy mileage because the engine overworks.

Yet, it's true that big, oversize engines deliver poor gas mileage, too. So, whether you buy a large car or small, get an engine that neither over- nor under-powers it. Break the engine in properly. Professionals "whip" a new engine during break-in. They speed up and slow down, never running at a constant speed more than a few minutes at a time. A thousand miles of highway-speed whipping, being careful that no overheating develops, does wonders for a new engine.

This chapter shows more ways to extract the most miles from whatever gasoline you get, in whatever car you drive.

Inside your car engine, the carburetor is the heart of fuel usage. It mixes air and gasoline to make the fuel vapor your engine runs on.

One adjustment controls *idling speed.* A spring closes the foot throttle (accelerator pedal) against an adjustable screw-stop, limiting how much gasoline/air vapor reaches the engine. Too fast a setting wastes a gallon or more of gasoline a week. You can usually adjust idling speed yourself if you have a screwdriver, nutdriver, or wrench that fits. You may have to lift off the air cleaner.

Idling speed is best set with a tachometer. With transmission in Neutral, and air conditioner off, set idling speed at 450 rpm. With an automatic transmission, raise idling speed to 500 rpm to prevent stalling at stop lights.

Rough idling may signal need for a tuneup. Or, *mixture* screws at the base of the carburetor may be maladjusted, upsetting the ratio of air to gasoline. Tromp the gas pedal and watch the exhaust. If it turns black, the mixture is too rich—not enough air. If the engine idles rough and hesitates (or backfires) when you accelerate suddenly, the vapor mixture probably is too lean—not enough gasoline.

A thorough tuneup technician sets carburetor adjustments while monitoring an electronic engine analyzer. The carburetor plays a significant role in your gasoline savings. Have it checked and readjusted twice a year, in spring and fall.

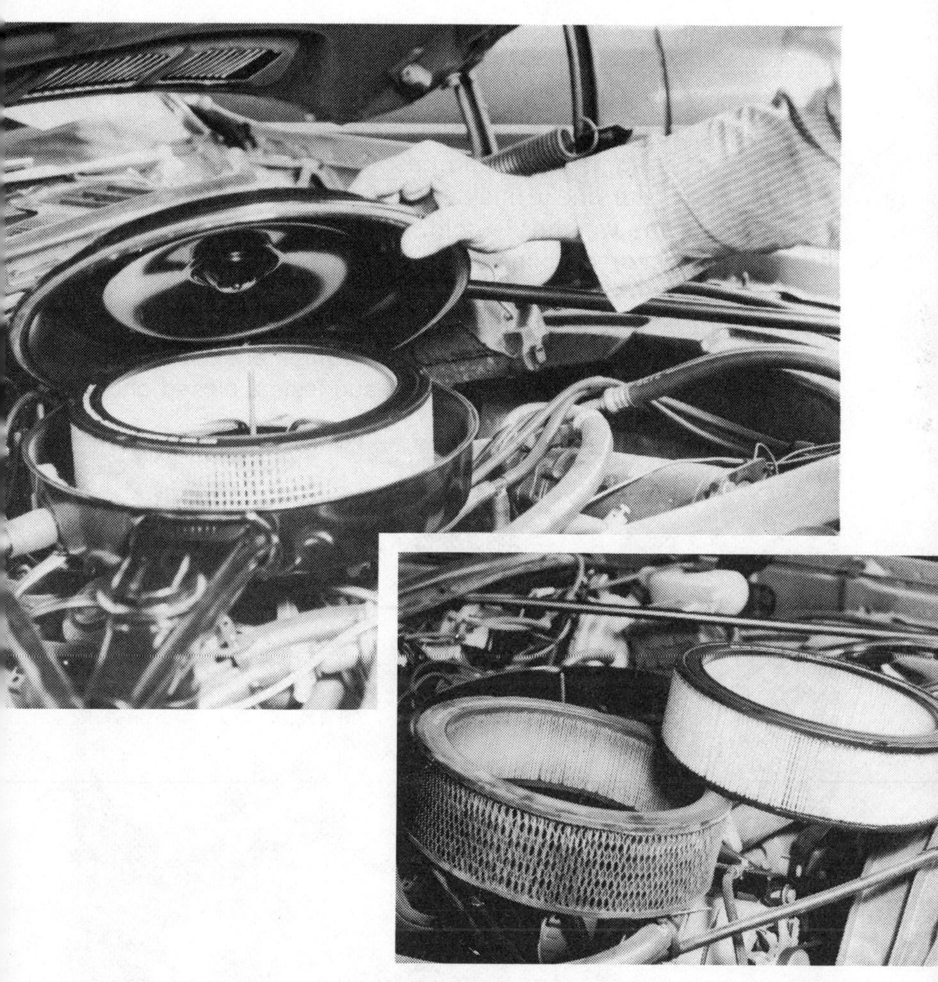

The *air cleaner* affects gas mileage. All air for the carburetor must pass through its filter to remove dust and particles. A stopped-up filter element starves the carburetor for air. An overrich and inefficient vapor mixture goes to the engine. Gas mileage goes down and driving costs go up. A dirty filter also lets junk into your engine that can do serious cylinder and valve damage.

You should inspect the air-cleaner filter every month. Unscrew the wingnut that holds the cover on. Lift out the element and hold it up to a light. If you have difficulty seeing through it, buy a new one.

Underneath the air cleaner, on one side of the carburetor, you'll find the *automatic choke* mechanism. Working properly, this device helps you start and drive your car economically. However, a sluggish or sticky automatic choke can be a constant source of aggravation and gasoline expense.

Stuck open, the choke makes your car hard to start when the engine or the weather is cold. If a stuck mechanism holds the choke closed, gasoline consumption rises drastically. A car that delivers 12 or 14 miles for a gallon of gasoline may drop to 6 or 8. The engine may also be hard to start; the carburetor "floods" easily. Black smoke from the exhaust pipe, even after the engine has warmed up, suggests a closed choke.

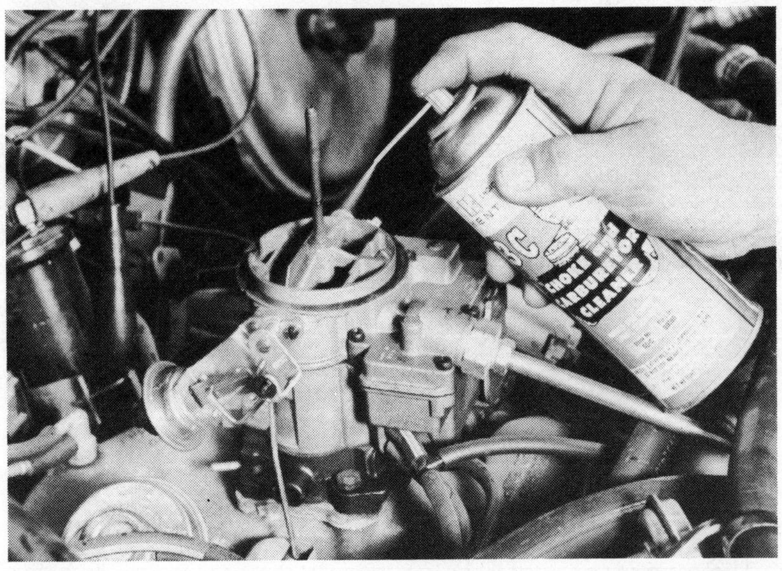

Try spraying the choke linkages with cleaner. Two or three sprayings usually clears off the gunk and frees the assembly. Or, use gasoline and a toothbrush to scrub all accumulation off the linkages. DO NOT oil them; that only attracts dust and assures sticky joints.

Your tuneup technician can disassemble the linkages and clean them, in severe cases. While he's at it, have him check the choke's adjustments. A car that runs on fast idle longer than is necessary after starting wastes a lot of gasoline on winter mornings.

The ignition system also affects engine performance, and therefore gasoline consumption. A hot spark ignites gasoline vapor inside each cylinder. The spark must occur at exactly the right time and in a form that burns the vapor (releases energy) at a precise rate. The *spark plug* is the gadget directly responsible for igniting the gasoline mixture.

Spark plugs can fail in many ways. Common faults develop at the electrodes on the bottom—right where the gasoline burns. An experienced mechanic spots certain engine troubles from how the plug electrodes look. Burnt-off electrodes (photo above) are the most obvious defect.

Any spark-plug fault can steal several miles from each gallon of gasoline you buy. Plugs should be inspected every 10,000 miles. A new set every 20,000 miles pays off in car performance and in gasoline savings.

If you change spark plugs yourself, make sure you buy the correct grade: cold plugs for sustained highway speeds, and hot plugs for stop-and-go city driving. And be sure you set the electrode gap correctly for your car—usually at 35 thousandths of an inch (.035″).

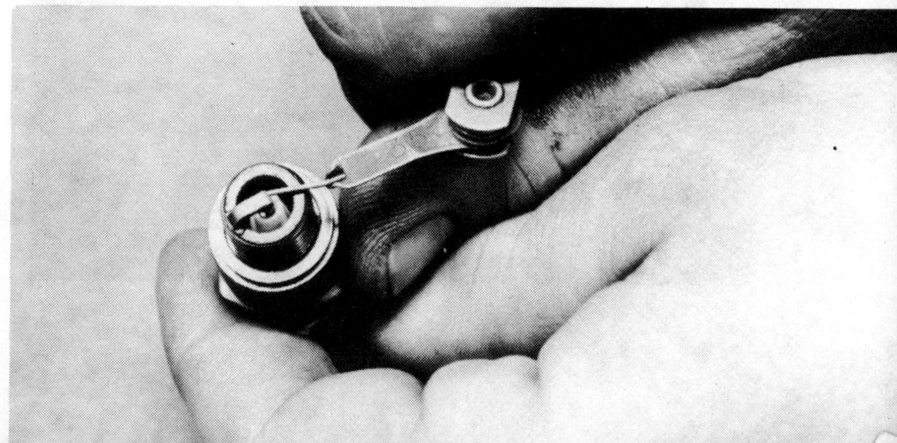

A second important link in the ignition system is the *breaker points*—inside the distributor. A few new cars have electronic systems that either eliminate breaker points or reduce sparking across them so they wear longer. Ordinary engines go through a set of points in 20,000 miles or so. The contact surfaces pit and play havoc with gas mileage.

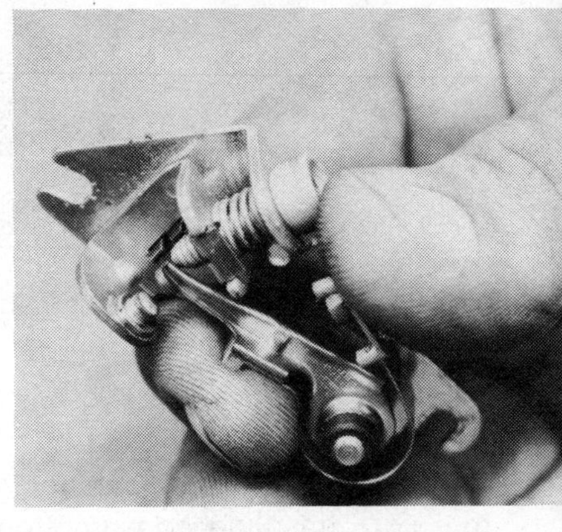

The cure is a new set. Any thorough tuneup calls for new breaker points, new spark plugs, a new rotor in the distributor, and often a new distributor cap. The technician should also test the ignition harness for leakage or erratic firing. He should check ignition timing and duration. And of course he cleans and adjusts the carburetor.

Careless tuneup shortens the life of engine parts. For example, a wrong setting of the breaker-point gap lets the contact surfaces break down far sooner than normal. High-quality points, adjusted critically with a dwell tachometer, and if engine timing is precise, could last 30,000 or 40,000 miles. Improved engine performance also saves you many gallons of gasoline in that time.

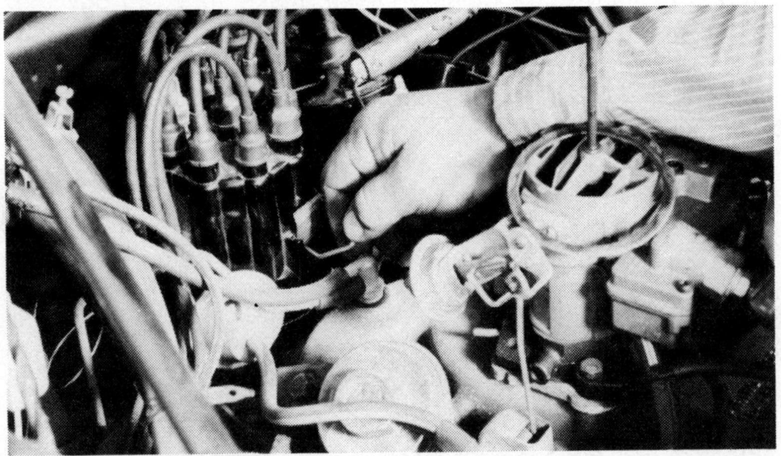

Cars with automatic transmissions have a reputation for delivering fewer miles-per-gallon than those with gear shifts. With most drivers, that's typical. But the sharp driver does just as well with an automatic. In perverse driving situations, the automatic transmission can even help save gasoline.

Next to how you drive, the transmission itself weighs heavily on gasoline consumption. Keep transmission fluid at proper level. You check it with a dipstick at the rear of the engine, usually on the side opposite the oil dipstick. After the engine is thoroughly warmed up, set the parking brake and put the transmission in Drive. (For greater safety, have someone hold the foot brake too.) If the dipstick shows ADD, put in a pint. When you're through, seat the dipstick cap firmly. Every 25,000 miles, have the fluid changed and a new sump filter installed.

Meanwhile, pay attention to how the transmission shifts. Whether upshifting or downshifting, the transmission should move smoothly from one gear to the next, without "chunking" or "breaking away" (letting the engine run free for a moment). It should not "hold" in any intermediate gear unless you are applying gas-pedal pressure. If you spot any of these symptoms, take the car to a transmission mechanic you can trust for adjustment. Besides wasting gasoline, these symptoms lead quickly to more serious transmission trouble.

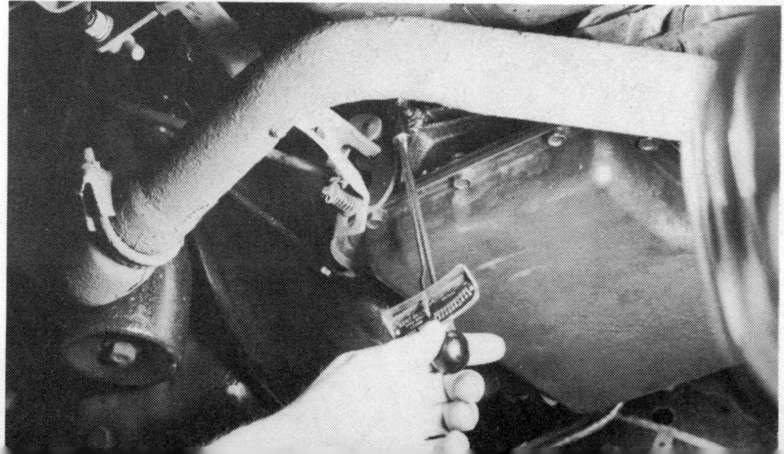

Top gasoline mileage depends on the condition of your entire automobile. The engine should have a minor tuneup every 10,000 miles or twice a year. A major tuneup can wait as long as 25,000 miles.

Check engine oil *every* time you buy gasoline. You never know what might have happened in the last couple-hundred miles. Don't buy cheap oil; it's no bargain to gas mileage or engine wear. Use 10W-40 grade for most driving. Check the owner's manual for your car, if either driving or weather is extreme. Change oil and the oil filter at least every 5000 miles. If the oil dipstick shows dirt and sludge, change sooner.

Keep the gas tank nearly full; there's less evaporation loss. But on warm days, don't fill up in the morning and let the car sit. Expansion causes overflow—spilling as much as a gallon on the ground.

Keep the chassis lubricated. Don't let your service attendant overlook the differential (photo below) or the manual transmission gearbox. Either can go dry without attention, wasting engine horsepower (and thus gasoline) and bringing on early repair bills.

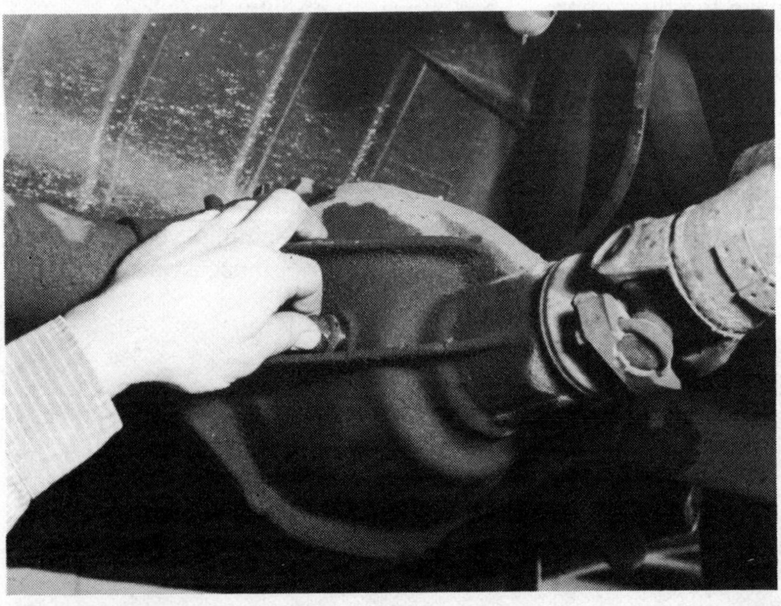

Many drivers waste gallons of gasoline right in their driveways. They don't know how to start an engine economically. Here's a two-part sequence that starts any but the most cantankerous engine.

(1) Buckle your seat belt. Put the key into the ignition lock. Pump the accelerator one time—not twice, but once. Turn the key to start. Do not pump the accelerator while the engine turns over, *unless* it doesn't catch on the third or fourth turn. In that event, pump the accelerator once more *while* the engine turns. Avoid bouncing the gas pedal for starting. An engine that takes that much pumping needs attention.

(2) If the engine turns ten or twenty times without starting, and you've pumped the accelerator perhaps three or four times, stop grinding. Wait eight or ten seconds. Then put the accelerator slowly all the way to the floor and *hold it there.* Try the starter again, *keeping* the accelerator down. Don't even pump once.

With a manual choke, pull the choke all the way out for the first several turns of the engine. Pump the accelerator once or twice as the engine turns. If no luck, try again with the accelerator down and the choke only halfway out. Whenever the engine catches, hold the accelerator one-quarter down and adjust the choke for smooth running. After three or four minutes, remember to push the choke all the way in.

Drive the first five miles under 30 mph, to let everything loosen up and get rolling freely.

Racing drivers know that two factors cut deeply into their fuel mileage. One is *wind resistance;* the other is *rolling resistance.* It takes horsepower to overcome these resistances, and expended horsepower means burned gasoline.

In your automobile, you combat wind resistance mainly with slower driving. The speed of your car incurs more wind resistance than any natural wind you're likely to buck. That's why the government rulings about 55 mph speeds (page 21).

Rolling resistance is more subtle. You reduce it partly by keeping the car well lubricated. Vehicles that sag forward or rearward impose heavy rolling resistance; helper springs or heavy-duty shock absorbers compensate if there's extra load aboard.

Tires, in their moving contact with the ground, contribute the most rolling resistance. That's why you should keep wheels aligned and tires well inflated. Just one slightly soft tire costs you gallon after gallon of gasoline on a trip.

Actual pressure depends on your load. If your car is light, use recommended tire pressures—generally 24 lb in front and 26 in the rear. But when your trunk is loaded or you have family in the back seat, those rear tires need more pressure. You'll get best gas mileage—and far longer tire wear, although the ride may not be as soft—with 30 lb in the rear and 26 to 28 in front.

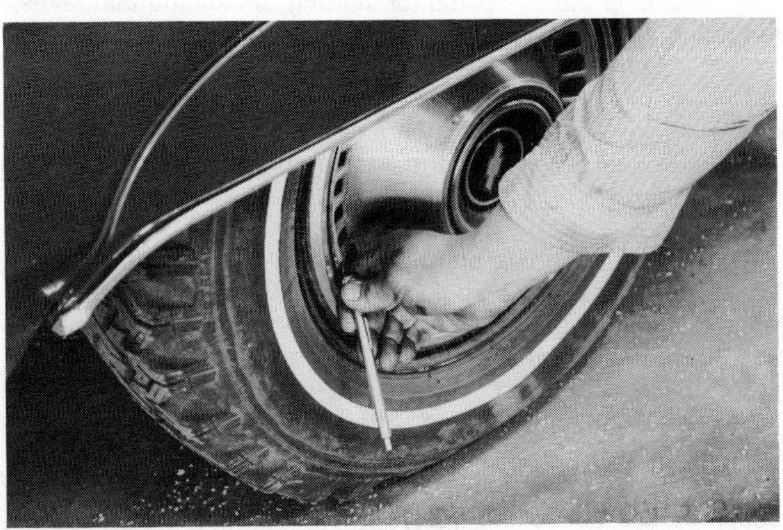

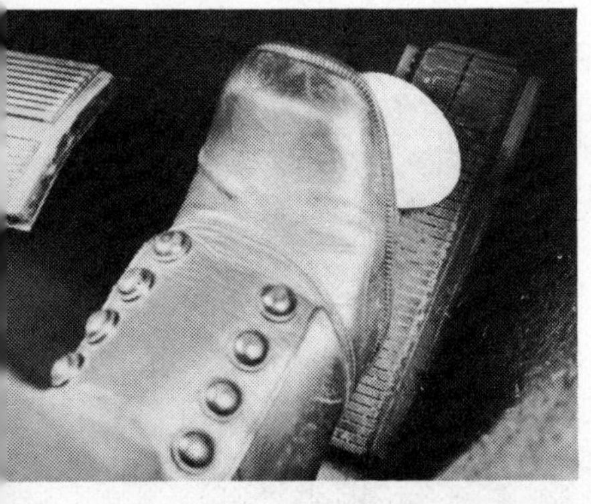

Professional drivers agree on one axiom: *The easy driver gets the best gas mileage.* Nothing you can do saves more gasoline dollars than treading featherfooted on that gas pedal. Experts suggest "eggshell" driving. You imagine a fresh egg between your foot and the accelerator. To avoid breaking the eggshell, you press gently and evenly.

This doesn't mean you drive at a snail's pace. Economy from slower driving ceases below 35–40 mph, as rolling and wind resistances become negligible. Tests show that slow drivers waste as much gasoline as speeders. What you do is treat the foot throttle delicately at all times. Going up hills, you don't tromp down on the gas. You plan ahead and apply only enough extra power to maintain speed without slowing down. In level driving, you avoid any pumping of the accelerator.

Literally, you can squeeze from 2 to 6 extra miles from every gallon of gasoline in your tank. That's a 25% to 50% saving in fuel expense. It's worth the extra concentration.

Getting any car under way takes horsepower and energy. Gasoline furnishes the energy. Don't waste that energy unnecessarily; it costs money.

The wipeout getaway may build up your ego, but it tears down your bank account. Leaving in a cloud of dust or a squeal of tires appeals only to insecure, showoff drivers anyway. A mere half-dozen such starts can guzzle a gallon of gasoline. That's not to mention the damage it does to tires, chassis, engine, and drive train.

Instead, move out intelligently. Put the automatic transmission in Drive and accelerate slowly. Entering a fast-traffic street or pulling away from a stop light, push lightly at first. Add power by pushing slightly harder as the car begins to build up speed.

Want to move out like a pro? Set a glass of water on the dash. It must neither slide nor spill a drop as you accelerate from zero to 60 mph. Any competent driver can make a habit of this smoothness.

Heedless and unnecessary braking wastes gasoline. How? It takes a definite amount of gasoline-derived energy to get your car moving at a certain speed. Inertia would keep it rolling for a great distance. Instead, jackrabbit stopping dissipates that accumulated energy suddenly. The stopping isn't the worst cost—just some needless brake wear. But *regaining* the speed does cost, in gasoline consumption.

You, as a bright driver, build up that rolling energy in your car steadily and economically by smooth acceleration. Then you get your foot off the gas pedal long before you reach your next stop. Momentum lets you "coast" against engine backpressure. The engine burns hardly any gasoline during the entire roll. Too, the backpressure slows you down without a waste of brake energy. You make a smooth, quiet stop.

How does a pro judge a smooth stop? If your passengers rock forward, that's too abrupt. You've wasted some of that expensively gained forward-rolling energy.

City driving can go heavy on gasoline, particularly with automatic transmissions. Most of the gasoline waste, however, comes from poor driving.

Look and think ahead. Watch stoplight patterns. Pace yourself to get to each one after it's green. If a speed is posted, assume the lights are synchronized to that.

Move *with* traffic. Cutting in and out usually means speeding up and then braking. Wasteful. Keep back a bit from traffic you're following. You can see better and aren't bound by someone else's stop-start driving habit. And don't "ride" one foot on the brake; it wastes power and wears out brake shoes.

Find routes that don't have so many stop signs and lights. The less starting and stopping you do, the less gasoline you burn.

If you get caught in a traffic jam, shut the engine off. Idling more than a minute in one spot wastes gasoline and hurts the engine other ways. (But watch your battery after a dozen or two restarts. Let the engine run fast a bit for recharging.)

Cram as many errands as possible into each trip; it costs extra to warm up a cold engine. Get groceries as you pick up the kids from school. Mail letters on the way to bowling. Share your chauffeuring with other parents.

Expressway driving takes less gasoline per mile than any other, but only if you play the game right. The worst gas-waster on the open road is speed. Remember wind resistance and rolling resistance? Both rise drastically as speed mounts.

One factor is the engine load curve. Put simply: engines in today's cars, with ordinary gear and drive ratios, work most economically between 45 and 55 mph. Below 35 mph, though wind and rolling resistances are low, engine efficiency is low too. Above 50 mph, efficiency falls off *and* driving resistances climb rapidly with speed; gasoline consumption per mile increases all out of proportion to what you might expect. That's why you find miles-per-gallon so much lower when you drive 70 mph than when you settle for the 55 mph advocated now by the government.

Plan your long trips explicitly before you leave. Allow for stops. Above all, don't lay out the course so you have to rush.

Save gasoline by entering and leaving an expressway correctly. *Never* stop on the entry ramp. Nor should you approach slowly and then put on a burst of speed. Instead, smoothly regulate your acceleration on the ramp until you're moving exactly with traffic flow in the first lane. Blending in is easy that way.

Exiting can be hazardous. Don't coast too far trying to save gas. Well-designed expressways have deceleration lanes near all exits. Let up on the gas only *after* you're in this lane. Tread lightly on the brakes. Let momentum carry you to wherever you have to stop or turn.

Driving a vehicle heavier than an automobile challenges the vacation traveler interested in gasoline economy. Here are some tips:

A hill? Accelerate—but not in a rush—to build up momentum. Once into the climb, don't push down hard on the accelerator in an attempt to add speed. If the vehicle slows too much, just relax and let the engine do its work in a lower gear without being forced.

In hilly country, an automatic transmission that's properly adjusted can do a more economical job than you can. The idea is to anticipate how much power the hill requires, and set your foot throttle accordingly. The transmission and engine handle the rest.

Camper trucks and motor homes (RV's) with standard transmission present only a mild problem. Well-chosen gear-shifting patterns add many miles to a tank of gasoline. These three principles work for professional truck drivers:

(1) *Keep engine speed constant* by choosing the proper gears and the right time to shift for the hill you're on.

(2) *Choose the engine speed* that works most efficiently. You'll recognize it because the engine neither labors nor races.

(3) *Keep the RPMs up.* In other words, don't let the engine lug down on hills. Shift to a lower gear while engine speed is still high, but starting to show signs of loading. As the grade levels out, don't let the engine race; shift back to higher gears.

Drive ahead. Plan stopping so you don't waste momentum. Anticipate hills and curves. You'll get a lot more miles-per-gallon.

Dragging a trailer cuts miles-per-gallon to the bone. You can lose 50% without batting an eye. That makes any trip cost double. You can cut the loss to perhaps 25% with a few common-sense practices.

The hitch must let the unloaded trailer sit level at the car bumper. A trailer that tips forward or backward handles awkwardly and chops gas mileage.

Pressure the trailer tires (cold) to 35 lb if the rims are drop-center, to 50 lb if they're two-piece (truck tires). Pressure the rear tires on your car at least to 30 lb.

Load so weight is balanced evenly from side to side, and is centered slightly forward of the trailer axle. Weight on the hitch should amount to 10% of the total (trailer plus load). Fix the load so it can't shift.

Experts find that trailers handle most safely and economically at speeds not exceeding 40 mph. Experiment to find your engine-load/performance peak. Try 40 mph for a couple of tankfuls. Figure miles-per-gallon. Then try 5 mph over and 5 mph under. The speed that lets your engine work most efficiently with that load improves gasoline mileage.

A stiff headwind or crosswind can double gasoline consumption. Consider keeping trailers and RV's off the highway when winds are high.

Inexperienced drivers with huge recreational vehicles and trailers pay dearest for their travel miles. The enormous weight demands horsepower to overcome rolling resistance. Size, shape, and speed affect wind resistance. Still, forethought and wise driving can dent the cost of operating a large rig.

Streamlined design can add a mile or two to every gallon of gasoline.

Don't attempt to pull a trailer with a car and engine too small to handle it. If you must push the gas pedal to the floorboard to maintain speed on expressway grades, your rig is under-powered.

Starting and stopping a lot wastes gasoline. Drivers who pull trailers economically use a variation of eggshell driving: pretend the trailer carries a load that shifts easily and drive so the load doesn't shift. Anyone who hauls horses develops this gentle-start, gentle-stop technique.

Avoid awkward, jerky gear changes on hills and winding turns. Choose the accelerator pressure that gives smoothest uphill performance with automatic or manual transmission.

Learn to back a trailer properly, and you have another way to conserve gasoline. It's wasteful to make a half-dozen tries at putting a trailer in a driveway straight. (If the trailer is light, consider unhitching and rolling it by hand.)

Launching a boat constitutes a common backing situation. On a steep ramp, you can waste enough gasoline to run the boat an hour or two. These pointers for boat-trailer backing apply to other trailers too.

Always pull in from a direction that lets you back the trailer to the left, not to the right. That way, you can watch conveniently.

Begin with car and trailer in a straight line. To start the trailer turn, twist the steering wheel to the right and begin backing. The front end of your car moves leftward, and the trailer begins a backward left turn. Immediately straighten out the steering, and let the car "follow" the trailer as it backs, turning.

When the trailer is *almost* aimed the direction you want, twist the car steering a little extra to the left. That swings the car's front end over straight with the trailer. Straighten the steering and continue backward.

Above all, don't hurry. Back slowly, watching the trailer's wheels. Correct any turning error instantly. Don't let it go so far you have to pull forward to prevent jackknifing. That's what costs.

Cars aren't the only machines that offer opportunities to save gasoline dollars. You should know the most economical ways to operate all gasoline-powered engines.

Boats, for example. Some gasoline-savings efforts are fairly obvious. Keep the engine tuned up. In outboard engines, check and clean spark plugs every 20 hours or so; install new ones every 100 hours. Put in new points each spring, and check them two or three times during the boating season.

Load the boat evenly. If you're the only passenger, arrange gasoline cans or other weight to keep the keel level. Keep weight forward so the boat planes quickly as you come up to speed.

Change props when you shift from pulling skiers to cruising. This lets the engine operate on the economical portion of its load curve.

Most important of all, pick a cruise speed 3–5 mph below top speed. That's a far more economical power setting for the throttle. By giving up 10% of speed, you can trim up to 25% off your gallons-per-hour gasoline consumption.

All-terrain vehicles (ATVs) and snowmobiles have taken the place of go-karts for weekend fun. All-terrain cycles (ATCs) and mini bikes are the rage among subteens, who aren't yet old enough for motorcycles.

There should be enough gasoline for fun and games, if you take the standard precautions against gasoline wastage. Keep the engine tuned up. A new spark plug every 20–25 hours should be regarded as standard maintenance. With four-cycle engines, be sure the oil reservoir stays full. Keep the oil/gasoline ratio in two-cycle engines carefully measured. Keep chassis lubricated. Clean air intake filters daily; dust and weeds extract a costly toll.

Some parents limit the gasoline committed to this kind of recreation. When a two- or five-gallon gas can has been emptied, the weekend "racing" is done.

You can even save gasoline with your lawnmower. Most important: keep the engine tuned up. New plugs and points every spring assure quick starting and lessen gasoline consumption. Keep the oil reservoir filled. If the carburetor has idling and load mixture adjustments, ask your lawnmower technician to be sure they're set for least gasoline consumption.

You can save a can of gasoline or two every season by mowing just a bit slower. Set the throttle wide open and then back it off just enough to noticeably slow the engine speed. That's an economical power setting (and incidentally results in more even mowing than top speed does). Go easy in tall grass; don't let the engine choke down. The newest gasoline power mowers have governors that pick out the best operating point on the engine's load curve automatically.

You may have realized as you've gone through this chapter, gasoline savings come down to two basic factors: (1) How well your machine and its engine are maintained. (2) How intelligently—and gently—you handle driving or operation. Attention to both will save you many dollars and will help alleviate any gasoline shortage that threatens.

Chapter 2

Coping With Electric Power Brownouts

Petroleum doesn't constitute the only energy threat in this country. Other fuel sources appear suddenly limited too. Certainly they're in short supply when you go to buy them.

The pinch affects powerhouses that generate electric energy. Almost everyone remembers the gigantic power blackout that struck the northeastern part of the U.S. in November of 1965. Power needs have risen steadily. Generating plants, particularly those dependent on petroleum and coal, can't keep up. The result has been brownouts—periods when power is insufficient to operate all appliances at normal voltages.

Unless demand for electric power quits growing so fast, brownouts may spread countrywide. This chapter suggests what you can do to avoid that kind of inconvenience—or worse, another total blackout. At the same time, you'll be lowering your electric bill.

The answer lies partly in each of us using less electricity. We needn't do without, just be wiser with what we do use.

Lighting offers many areas for saving electric energy. We use lighting to guard against accidents, to illuminate our work and play, to let us read and perform close work without headache, nervousness, or eyestrain, and to beautify our homes as well. We must somehow stretch lighting efficiency without giving up any of these.

Adequate incandescent lighting calls for about 1 watt per square foot, but not all in one spot. Large rooms—more than 250 sq ft—should have a fixture for every 125 square feet. In an average 200-sq ft bedroom, you'll want reading light near the bed. A 100-watt bulb there lets you spread the other 100 watts over the remaining area.

If incandescent light seems dim, consider fluorescent lighting. The tubes last longer than incandescent bulbs and give far more light per watt of electricity consumed. The popular cool-white fluorescents may give you a headache; try daylight or warm-white. The softer shade is compatible with general living quarters and bathroom, although you may prefer cool-white in kitchen and workshop.

Quick saver tips: Long-life bulbs burn longer, but deliver less light per watt. Turn off incandescent bulbs anytime you're not using them. Leave fluorescent lights on if you will return to the area (kitchen sink, bathroom) within a half-hour. Mercury vapor or sodium lamps may suit your purposes best outdoors.

Clean ceiling fixtures assure better light and longer life for the bulbs. Dirt traps heat. Fixtures and globes should be part of monthly housecleaning.

Use natural light when and where you can. But avoid placing a desk or chair where direct sunlight strikes your reading material. Nor should you face the glare of outdoors. During the day, especially for youngsters' homework, keep drapes or curtains open. Use only what supplemental lighting they (or you) need.

See that the correct size of bulb is in each fixture. Over size bulbs brighten a room but they overheat the fixture, shortening its life and that of the bulbs. For more light, add fixtures or portable lamps. The Better Light Better Sight Bureau recommends 200 watts of incandescent illumination for studying. Few study lamps will safely power a 200-watt bulb, so use two lamps. Little high-intensity lamps are bad for your eyes when used without any surrounding light. The contrast between light and dark areas is too much.

There are other ways to conserve electricity around home. For example, your appliances. Check how much electricity is required to power them. This is usually listed on the bottom of the appliance, in *watts* (the same as light bulbs).

Compare one brand to another. Wattage efficiency sometimes varies among models. Extra wattage may be due to additional functions. The object is to get done whatever work, cooking, grinding, etc., you need done, but using the least possible watts.

To dramatize wattage conservation, consider this example. A black-and-white tv set in average use consumes about 140 kilowatt-hours (kWh) less electricity each year than a color set. That 140 kilowatt-hour saving could power all these other items (in average use) for a year: radio/phonograph (109 kWh), hair dryer (14 kWh), sewing machine (11 kWh), vibrator (2 kWh), electric shaver (1.8 kWh), and electric toothbrushes for a family of four (0.5 × 4 = 2 kWh) for a grand total of 139.8 kilowatt-hours, leaving 0.2 kWh to spare.

It pays to figure out wattages. You can save a lot of money in electricity and you can maybe live better than you thought.

Speaking of tv sets . . . they have become virtually part of the family (most often the babysitter). What happens without tv? Mother misses soap operas, Junior and Sis grumble because they can't see cartoons and triple-rerun movies. Papa, after his long, hard day at work, can't relax after dinner with tv.

To avoid such traumas, learn to conserve your television set. Here are some hints.

Don't leave the television on all day to catch the soaps. Set the oven timer or an alarm clock to go off 5 minutes before your program airs and you won't miss anything. Turn the set off when your program is over.

When buying a new set, select a solid-state model. It uses about half as much electricity as similar tube models. Color sets use more energy than black and white.

Unplug the television when it's not in use for long periods.

Enjoy your television set, but don't waste it. You'll also be conserving precious electric energy.

The kitchen offers many chances to conserve energy. The next few pages contain simple suggestions that increase appliance efficiency while reducing electric consumption.

Let's begin at the refrigerator. Some are cold-storage garbage cans! Clean out whatever's beyond use (you might want to glance through Chapter 6 first) and straighten up what's left. Neat placement lets air flow freely around every item. That saves electricity. Use your refrigerator, but don't overfill it.

Proper storage in the freezer compartment is exactly the opposite of what's right for the refrigerator. Keep your freezer full. A filled freezer retains cold and hence uses less energy. The bulk of the frozen foods helps keep warm air from rushing in when the freezer door is opened.

Do not allow frost to accumulate more than ¼ inch. Defrosting at that point takes little time and foods won't have time to thaw at room temperature.

If frost collects often, or if water condenses in the refrigerator, check the door seal. If a dollar bill slips easily (see photo), the gasket is not functioning properly. Check all around the door, as one side may seal and another may not.

Keep the coils (underneath and/or behind) dust-free. Open the grille at the bottom and carefully dust with the crevice tool of a vacuum cleaner. Condenser coils on the back take the dusting attachment. Be gentle but thorough. Clean the coils monthly—when you wash out the interior with a solution of water, mild detergent, and baking soda.

Quick saver tips: Do not put warm dishes of food in refrigerator or freezer. Don't put refrigerator or freezer near a range, oven, or other heater. Block any direct sunlight. Double-door refrigerator-freezer units are more efficient than one-door combinations. Try not opening either door as often. Keep refrigerator temperature at 38° or 40° (use a thermometer).

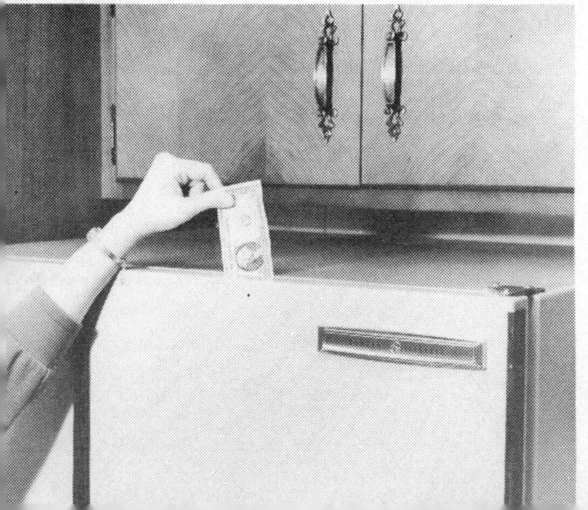

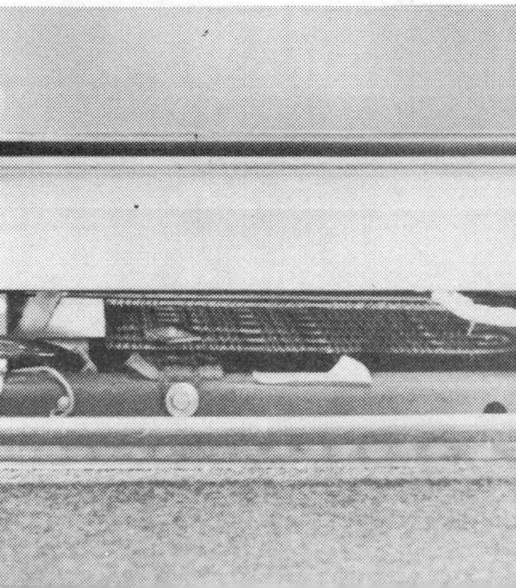

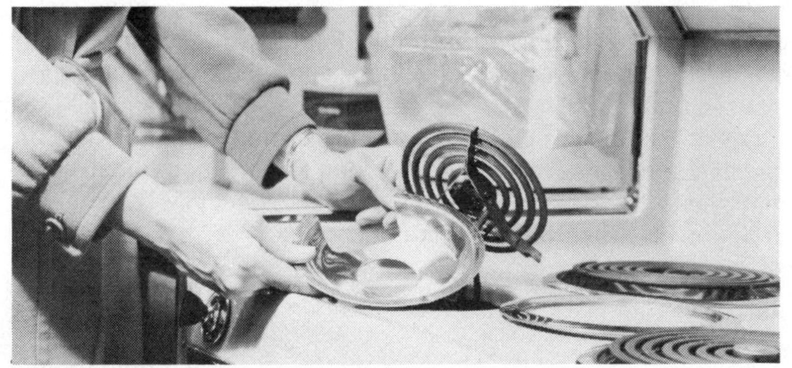

Electric ranges use from 8,000 to 16,000 watts. With that much to waste, planning for efficiency is worthwhile.

The key is maintenance and care. A clean range wastes less energy.

Reflector trays concentrate heat at the bottom of the utensil, where it does the most work. Food-encrusted reflectors waste heat. Clean them daily.

Do not wipe off a hot stove. Water on hot surfaces can damage both finish and function.

Some burners are thermostatically controlled. The control knob has degree markings. They save energy, if you know the best cooking temperature for the food on the burner.

Don't line the oven with aluminum foil; it hampers heat radiation and reduces oven efficiency. The oven operates better clean, too. After it has cooled, clean the oven each time you use it with a mild detergent-and-water solution.

Some ovens, particularly the self-cleaning type and ranges with microwave ovens, have a seal around the door. Check this seal frequently.

Clean the range hood and filter weekly, by washing the filter in a mild soapy solution and rinsing thoroughly.

Investigate the variety of cooking methods beginning on page 40. They are alternatives to range-top cooking, and they use less energy, time, and money.

You can save electrical energy in how you use your dishwasher too. For example, wash only full loads. Load the washer properly (see your owner's manual), and utilize the unit's special features.

Open the dishwasher and let the dishes dry naturally. If dishes are washed at night, they don't need those heat coils that speed the drying. Some models automatically bypass the drying cycle when the door is opened.

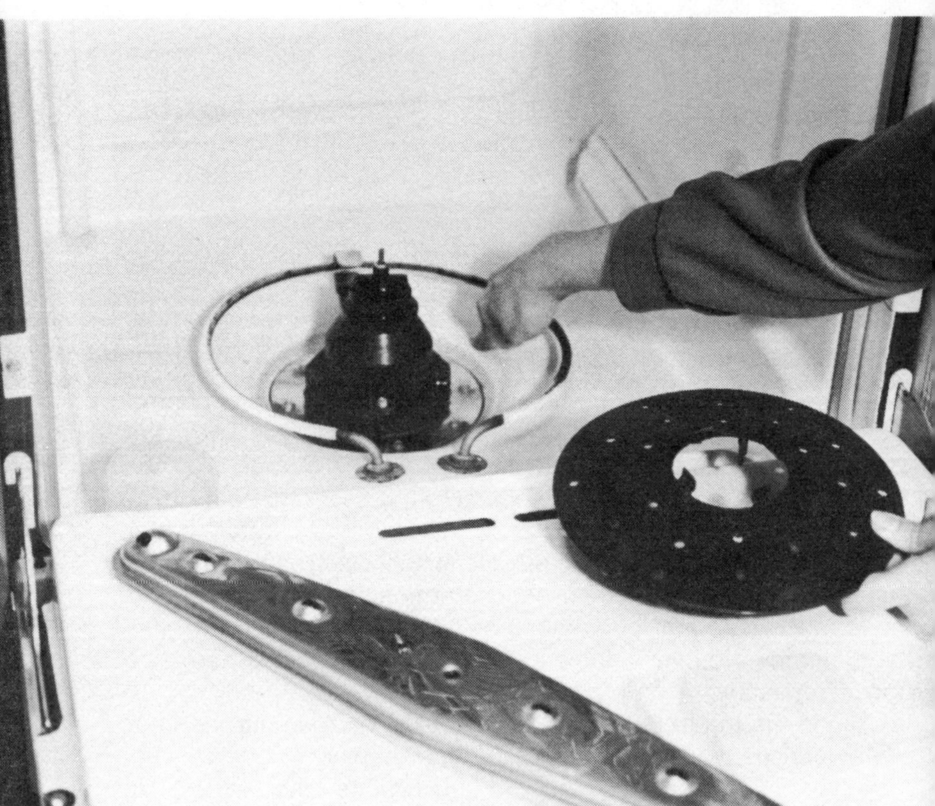

If your dishwasher doesn't have a built-in disposal unit, clean the strainer frequently. It's under the water propeller, which either lifts up or unscrews easily. The strainer screen lies beneath the dark protective shield.

A water heater requires from 2,000 to 5,000 watts, which makes it the third largest energy consumer in your home. Your dishwasher uses the hottest water needed. A temperature between 150° and 160° is sufficient. Without a dishwasher, use a setting around 140°.

A water heater should be centrally placed. Long pipes cool the water and thus waste energy. An ideal situation would put kitchen, laundry room, and baths all near the heater. Baths on other floors could be located directly above or below.

Deep tub baths use much hot water. Quick showers clean you as well. To conserve even more, first wet your body. Turn off the shower and soap. Turn it on again to rinse. That's quick, clean, and thrifty.

Leaky hot-water faucets cost you money. Fix them. Replacement faucet washers are cheap.

Some gas water heaters have a low vacation setting. If you will be away for several days, use it.

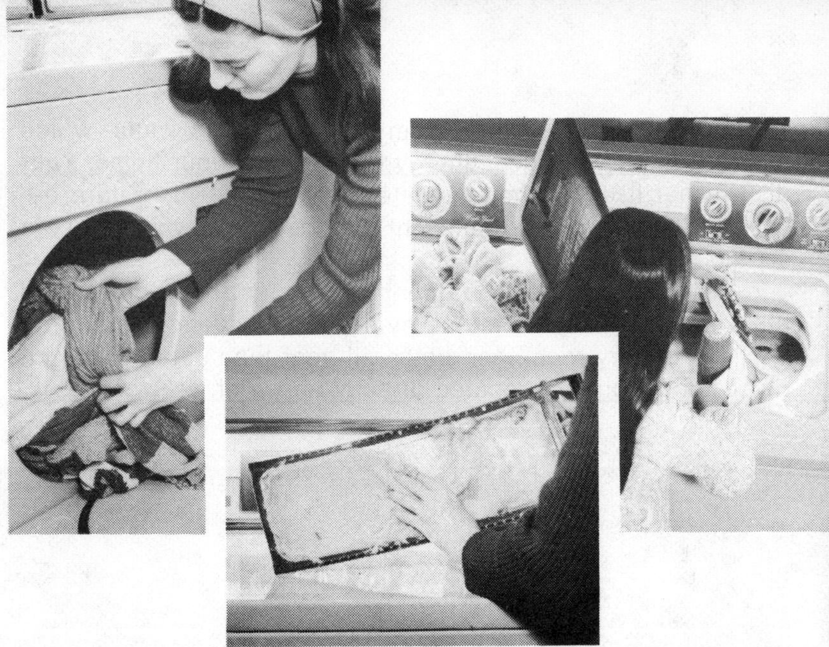

Laundry equipment consumes lots of electricity. Heating the water and running the dryer (5,000 watts) take the most energy. The washer requires 700 watts. Any efficiency you can contrive, you need.

With modern detergents, only your white cotton sheets need hot water. That saves some energy. You might want conditioners to break down hard water, which saps a detergent's effectiveness. A trick to this: As the washer fills with water, add conditioners. When filled, add detergent and let it all stand without agitation for about two minutes. Then add the clothes and resume the washing cycle. The additives thus have time to prepare or precondition the water for a more thorough wash.

Quick saver tips: Wash full loads. If your washer has a soak cycle, use it; work and play clothes come cleaner with less wear. Use water level controls, if your machine has them. Dry full loads, but don't overload the dryer. That causes clothes to wrinkle and wastes energy during ironing. Avoid overdrying; it makes clothes hard, wastes energy, and adds to ironing time. Use the lowest dryer temperature that will dry the clothes in a half-hour or an hour. Use Warm for permanent-press clothes and remove them before the drying cycle ends. Fold or hang immediately. Fluff-dry throw rugs. Clean the lint filter after every load. Check the outside vent and clean away accumulated lint.

Ironing can be a heavy chore. So, avoid needless ironing and you save yourself as you save electricity.

Schedule laundry so you have at least a full load to iron at a time. You don't waste electricity heating the iron up so often.

Select the lowest temperature and iron your delicates and synthetics first. Then raise temperature to accommodate the heavier fabrics. Never use higher temperatures than are absolutely necessary. Turn the iron off before you finish; the residual heat is fine for touch-ups.

Damp clothes iron far easier than hard, dry ones. Sprinkle clothes an hour before ironing, or damp-dry and iron immediately.

Keep your iron clean, especially the sole plate. Accumulations of starch, fabric finishes, or overheated synthetics can be washed off when the iron is thoroughly cool. You can't use abrasive cleansers, so get at it before the crust becomes thick.

Distilled water won't clog up your steam iron like tap water will. Lime deposits bring on steam-sputtering and may discolor the steam—which discolors clothes.

Replace a frayed cord with an asbestos-covered replacement. Don't buy any cord with smaller-diameter wires. It wastes power and could even cause a fire.

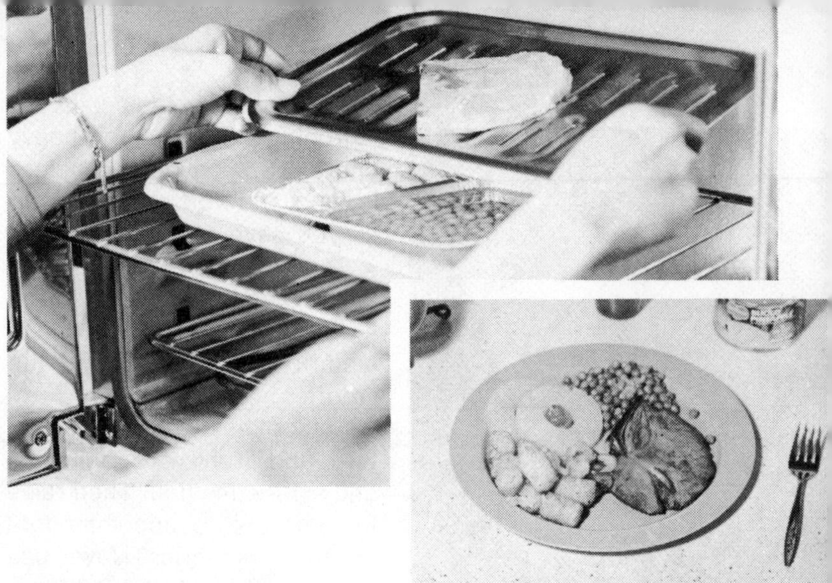

We promised, for the sake of heading off power brownouts, to show ways you can conserve electricity in meal preparation. Broiler meals are one example. They are an uncommon specialty, offering a chance to be creative.

You'll need a broiler tray with a grid to hold the meat so fat can drain away. The pan holds vegetables.

Build your meal around the meat. Select lean meat. If steaks, ham, or chops have fat around the edges, pare it off. Snip the meat at the edges to prevent curling. Smoked meats broil nicely, as do fish and fowl.

The meat takes longest to cook, so start it first. Thaw frozen vegetables at room temperature. Frozen vegetables, even pre-thawed, must cook longer than canned. You could place frozen vegetables in a shallow covered pan in the oven while the meat broils. Frozen tv-dinner trays make handy pans; they keep vegetable juices from mingling.

When the meat is ready to be turned, season it, then turn. That's also the time to uncover vegetables and put them under the grid in the broiling tray. Place all back in the broiler. Just before the meat is finished, you may wish to garnish with canned fruit such as peaches, pears, apricots, or pineapple, or fresh banana or apple slices.

Everything cooks at once (saves oven energy). Broiled meals like this leave fewer dishes to wash (saves dishwasher energy). Broiler meals get you out of the kitchen quickly (saves your energy).

One-pot cooking reduces electrical consumption. Using one burner for each hot dish consumes up to three times as much energy as one-pot cooking. One-pot meals can usually be prepared in electric deep-fryers, skillets, crock-pots, or electric woks. Crock-pots cook slowly. Electric woks and deep-fryers get it done more quickly.

The electric skillet, thermostatically controlled, suggests a wide variety of uses. According to Sunbeam researchers, it takes 30% more power to cook six 4-oz hamburgers on a range than in an electric skillet. Baking a cake in an oven consumes 60% more energy than doing it in an electric frypan. That's because of energy wasted heating that big oven; you use every ounce of space in the frypan.

A frypan cake tastes just like a baked one, but the shape is different. Use regular cake mix. The baking takes 30–50 minutes at a temperature setting between 260° and 300°. If the cake comes out too dry, lower the setting next time. Be sure to open the lid vent so steam can escape.

Preparing an entire meal in an electric pot is even easier than "frying" a cake. Stews, soups, and casseroles are all one-pot dishes.

Tasty meals can be prepared entirely in a pressure cooker. Try chicken with noodles sometime. Delicious—and far quicker than roasting or boiling the chicken first. Add butter or margarine for richer flavor. A pressure cooker on the range-top conserves energy by cooking quickly. Just keep an eye on pressure, and NEVER attempt to open a pressure cooker before you have bled off the pressure.

Many recipes cooked in crock pots take eight hours to finish. This is good for less expensive cuts of meat; the slow cooking tenderizes the meat without shrinking it.

Want your meat and potatoes in a big hurry? Try a microwave oven. This appliance doesn't really bake, in the usual sense. The microwave energy scrambles the molecules in the food to such a high state of activity that they heat up and cook. Most foods thus cook from the inside out. They don't brown on the outside unless an infrared browning element is included. Microwave ovens cook up to 75% faster than other ovens. A 4-lb roast can cook in 22 minutes, hamburgers in 60 seconds, hot cereal in 1¼ minutes, frozen corn in 5 minutes, a baked potato in 4 minutes.

Microwave ovens consume more electricity (watts) per second than standard ovens do. However, since they cook so ex-

tremely fast, microwave ovens conserve energy. It's also a more efficient kind of cooking energy in that it does not heat up the utensils (never use metal) or the oven itself. Metal utensils can cause arcing around the magnetron that supplies the microwave energy. Even if that doesn't burn out the costly magnetron, it wastes a huge spurt of electric energy.

If you want to cook a tv dinner, remove the contents from the aluminum-foil tray and place in glass, plastic, or paper dishes.

Energy leakage around the door is critical with a microwave oven, as it might damage body tissues. Safety interlocks aim to prevent leakage. Some brands have as many as three locking systems. If the door seal gets damaged, or when the oven becomes a little aged, have the seal checked and/or replaced.

Keep a small plastic cup of water in your microwave oven whenever you're not cooking. Then if the oven is turned on accidentally, the water absorbs energy which might otherwise harm the device. Use your microwave oven efficiently, to prepare entire meals. Foods prepared by microwave retain almost all their natural juices, and are very flavorful. So enjoy.

Cakes, cookies, and other pastries fresh from the oven can be delightful. And the aroma of Christmas goose or Thanksgiving turkey is absolutely mouth-watering. These are a mere beginning. Use your oven often. It can be economical and energy-saving.

Like with one-pot cooking, your oven allows you to cook a whole meal at one time in one appliance. In fact, with reclaimed tv-dinner trays an oven meal can be multiplied into several tasty side- and main-course dishes, or a variety of home-cooked tv-style dinners. Cook more than you need for one meal and save hours of cooking time later.

When dishes to be frozen are done, remove them from the oven and cool. Never place hot dishes in your freezer.

When you use your oven for several foods, choose foods that cook at about the same temperature. But you must know how long each item requires. Meat usually takes longest. At the appropriate times, add side dishes. Everything should finish at the same time.

Almost anything that takes the same temperature range can be cooked in the oven together. Empty space wastes energy. Pies, for example, can be baked right along with other things. Refrain from opening the oven door any oftener than necessary.

These three pages are devoted to quick saver tips for conserving energy in your cooking.

Use flat-bottom pans. Warped pans do not make solid contact with the burner, and thus waste energy. No, all pans do not have flat bottoms. The bottom may warp when you place an empty pan on a hot burner or run cold water into a hot pan. To check for pan flatness, place a straightedge (ruler or such) across the bottom of the pan. If you see space under the ruler at either edge or in the middle, the pan is warped.

Choose pans that fit your range's burners. Undersize pans let heat escape and oversize ones cook unevenly; both waste heat energy.

A pan should be only about one-third larger than its contents; heat stays concentrated better. Cover all pans with lids to reduce heat waste.

Use only enough water in a pan to create steam. Few items require more than a cup of water. Most need less than half that amount.

Don't overcook foods. Vegetables should be cooked only until tender. Meat overdone gets tough and loses flavor. An exception is pork; always give it plenty of cooking time and temperature.

If you need to boil water on the range, start with hot tap water. Place pots or pans on burners before you turn the heat on. Turn burners off just before food is done. Heat left in the coils will finish the cooking.

Don't preheat your oven except for delicate baked goods. A glass window in the oven doors lets you keep it closed while baking. Turn on the oven light to check progress.

Thaw frozen foods before you start them cooking. Canned foods cook more quickly than frozen, and may be slightly less costly; compare prices.

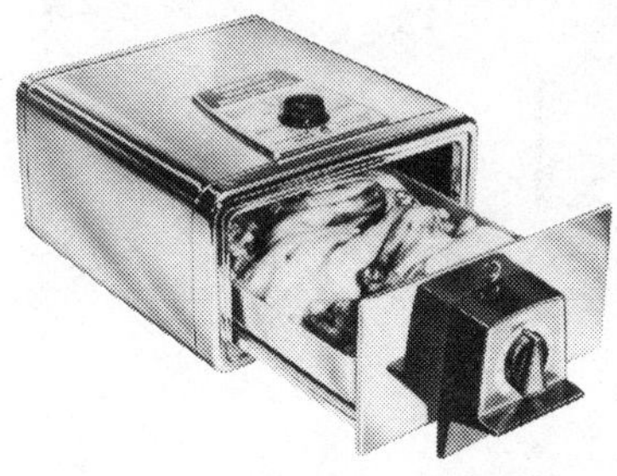

Investigate small cooking appliances, such as the portable speed oven, for tabletop cooking. Or find an electric blender that heats as well as purees. There are several new cooking appliances designed with energy conservation in mind.

Portable roasters can be used to cook entire meals at once. They require less energy than the range oven, because they're so compact. Many portables have timers, self-basting lids, and a thermostat control.

If you have a rotisserie, use it. It cooks meat about 30% quicker than conventional oven-roasting. Rotisserie-prepared foods are basted in their own juices, so are usually very tasty.

Never throw away your owner's manuals. Put them all together where you can find one when it's needed. Today's manuals give you great energy-saving tips on using the appliance,
and also instructions for minor repairs.

If you drop an appliance, check it immediately. Some damaged appliances can be repaired, others can be replaced at
less cost.

Turn off appliances that are not in use. Heading off brownouts
consists mostly of good-sense use of electric power. Don't
waste, and there will be plenty of energy for everyone.

Chapter 3

Getting Comfort Out of Heating/Cooling Fuels

Umpteen million U.S. homes are heated with fuel oil, a petroleum distillate similar to kerosene. Furthermore, certain utilities depend on a heavy grade of oil to fire boilers that drive steam turbines to generate electricity. So, whether you heat with oil or electricity, a shortage of crude petroleum can ultimately crimp your heating.

Gas users fare somewhat better so far, yet shortages in parts of the country are imminent. The point is this: unless we use heating and cooling judiciously, some homes may be uncomfortable part of the year. This chapter explains how to make the most of heating and cooling equipment, no matter what kind of energy does the work. The object is comfort, while saving fuel energy and dollars.

The heart of heating control in your home probably is a thermostat. Theoretically, it maintains a fairly constant temperature throughout your house. In actuality, it senses air temperature in the room where it's mounted. It turns a heating system on when temperature in the room drops, and off when the temperature climbs back to normal. If the heating duct system has been properly designed, a thermostat in one room can keep the house rather evenly comfortable.

You'll experience some "lag." That is, temperature falls a couple of degrees below normal before the thermostat turns on the heat, and then rises a bit above normal before the heat goes off. If the lag exceeds those few degrees, you'll get chilly between-times.

Federal recommendations say you should keep your home thermostat set for a winter room temperature of 68° in daytime. As you'll see in later chapters, that can be plenty warm or rather chilly, depending on what you're doing. However, avoid running the thermostat up and down often; that wastes fuel. Let the temperature stabilize for four hours or more at a time.

At night, the cooler you set the temperature for sleeping, the more fuel and dollars you save. A temperature average of 65° suits most people. Yet, 60° is just as healthy and feels cozy if you sleep under a few blankets or quilts. Setting the thermostat that low, if you leave it there for 6 hours or more, can trim about $15 out of every $100 you've been spending for heating.

Forced-air heating dominates, whether fueled by oil, natural or lp gas, or electricity. Yet, thousands of older homes—and many offices—use hot water or steam circulating through pipes to heavy radiators. Temperatures in separate rooms can be regulated by adjusting how much steam or hot water flows in each radiator. A few, usually in offices, have an automatic control on each radiator.

A thermostat usually controls the boiler where the water or steam is heated (usually by gas or oil). It takes some experimenting with radiator settings to equalize temperature in several rooms of a house.

The easiest technique is to turn off (clockwise) the valves of all radiators except in the room with the thermostat. If possible, close that room off. Let the temperature there stabilize. This takes a little time, because steam and hot-water heating systems react slowly.

Then, one at a time, begin opening the radiator in each room. It may take an hour for the effects of each one to become settled in its particular room. Once they're all regulated for even temperature throughout the house, let the thermostat take over. Don't monkey with the valve knobs any more. Setting the steam or hot water system like this assures even and economical heating the rest of the season.

With hot-air heating, the secret of keeping a whole house comfortable with one thermostat lies in the heating ducts and registers. They offer two facets of control.

Dampers determine how much hot air the blower pushes through each particular duct. Example: Hot air rises. Your upstairs may tend, on cold days, to be uncomfortably warmer than the thermostat keeps your downstairs. The trick is to close the damper partially in the duct leading to the upstairs registers. (Or, close the registers.) You are not wasting heat by doing so. You're merely proportioning the total heat differently—more for downstairs where the cooler air settles. A bit of juggling how much hot air reaches each room can even out the temperatures throughout a house. Result: more comfort at less cost.

Occasionally you find that distant room which simply doesn't get enough heat. The culprit, usually, is a duct that is somehow blocked off. Or, it may run through an unheated crawl-space, which cools the forced air before it reaches the room. A heating contractor can find the blockage or insulate the duct for you (or you can).

Clean hot-air registers regularly. Dust and dirt cut down efficiency. Keep cold-air return vents unblocked (more about this on page 77).

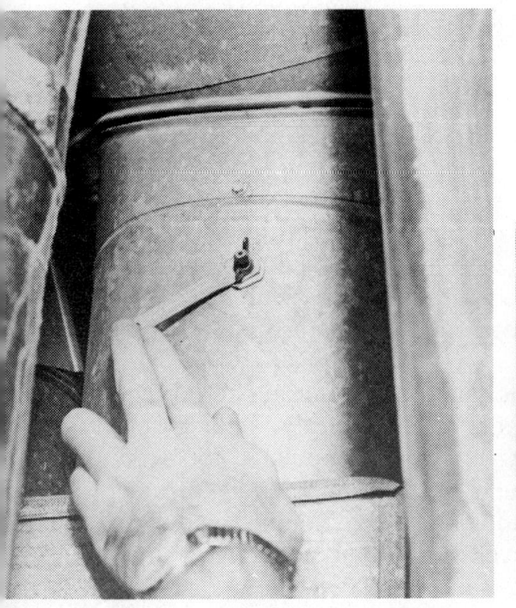

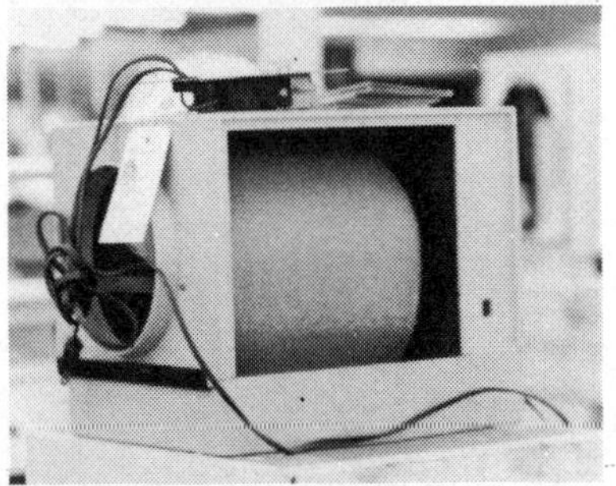

COMFORT is the name of the heating game. Temperature, as you'll see, is merely relative to other factors in your indoor environment. If humidity in a room is too low, which it often becomes in wintertime with heat on, your body gives up moisture too readily—and that cools you without your even being aware. Keep the humidity up to 60% or so and your comfort factor improves without any increase in heat. But don't raise humidity enough to cause sweaty walls or mildew.

Small humidifiers take care of dryness in a room or two. An in-the-duct automatic humidifier can keep the whole house comfortable. This unit injects a measured amount of vapor into the hot-air ducts from your furnace system. It can even be controlled by a *humidistat,* the humidifying equivalent of a thermostat.

Of course, in the summertime, when the central air conditioning turns on, you disable the humidifier. That's because hot-weather comfort depends on the air conditioner performing as a dehumidifier. You don't want the two fighting each other and thus wasting energy.

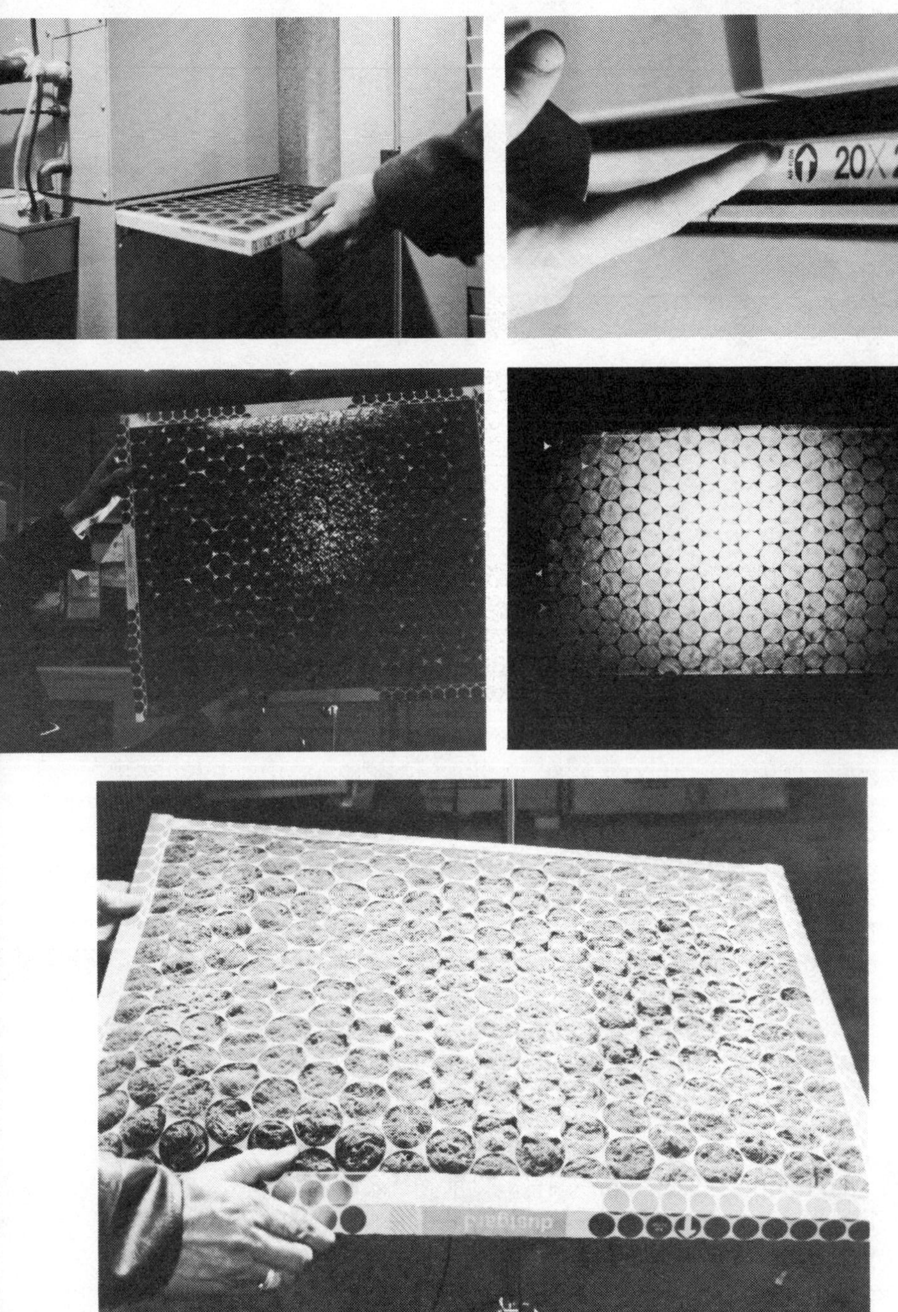

Blowing heated air through ducts ranks today as the most common home heating method. The air may be heated by oil burners, by natural or liquified petroleum (lp) gas, or by electric heating coils. Whichever way your furnace is "fired," principles of operation are essentially similar. A thermostat turns on the heating elements when house temperature drops. Heating elements raise the temperature of air inside the furnace shroud. An internal control then turns on the circulating blower. That forces hot air along ducts all over the house. When room or house temperature reaches normal, the thermostat cuts off the furnace heating elements. The blower continues awhile. That's to cool down internal temperatures in the furnace and take maximum benefit from all the heat that was generated.

Even when the same unit operates in summer as a central air conditioner, clean air to blow is a factor in efficiency. Dirty air clogs up the blower fan and motor, and fills ducts and registers with dusty grime deposits. Dirt can form an insulating layer over the heating or cooling chamber and prevent efficient transfer of heat to (or from, for cooling) the blower compartment. Result: poor efficiency and wasted fuel.

Every furnace or air conditioner has an air filter. You must inspect it regularly—never less than once a month.

If it's extremely dirty, you can see the accumulation of dust. However, dust collects down inside the filter too. Hold the filter between you and a strong light bulb. You can see the light easily through a new filter. A dirty filter blocks light. It also cuts down air flow when it gets this filled, wasting heat or cooling that cost you fuel-energy dollars to generate.

When you replace the filter, orient the arrow to point *toward* the blower (the air is filtered before the blower has a chance at it). Dirt can slip through if the filter faces the wrong way.

A fringe benefit of the energy crunch may be disappearance of the summer cold. Keeping indoor temperatures nearer to those outside causes less body change. The rules now say to set air conditioning at 78° in the summer. That still creates a cooling refuge from 90° or more outside, especially since the 78° air is dehumidified.

You can help your air conditioner consume less energy. Use an attic fan to draw away heat that accumulates between ceiling and roof. Keep drapes, blinds, and furniture from obstructing air flow. Keep lamps away from the thermostat; heat from the bulb could make the air conditioner run too much.

Inspect the unit regularly. Change the air filters. Dirt in the thermostat could cause a malfunction; remove the cover each spring and fall and carefully blow dust out. Brush any leaves and debris away from the condenser unit located outside. Trim grass and shrubs from around it.

A small room air conditioner, usually a window-type, can aid nighttime comfort, even with central air conditioning. The latter is probably still coping with heat accumulated during the day. That could leave bedrooms too warm for comfort. You could set the thermostat up 5° or turn the central air conditioner off, close the bedroom door, and push a small rug against the base of the bedroom door. Cooling one room this way is cheaper and more efficient than continuing to cool the whole house.

Individual room heaters (and portables) are great for warming special areas. Space heaters powered by gas or oil MUST be installed solidly and ventilated. Electric is better and safer for this purpose. Some space heaters have blower attachments to provide air flow. A separate fan, set at low speed, can also force air movement; keep it far enough away from the heater that it can't hamper operation.

Select a heater with a thermostat. A trip switch on a portable automatically turns the heater off in the event it's knocked over. Electric space heaters require more energy than most small appliances. Do not use an extension cord, or else buy an especially heavy-duty one.

In the interest of trimming fuel bills and conserving energy, you should know certain things about your heating system. These two pages, and the six that follow, discuss oil furnaces. Some of the principles mentioned apply equally to other kinds of furnaces, so read all of this chapter to find out the most about fuel economy.

A motor on the front of the furnace performs two functions. One, it turns a fan that creates a strong flow of air into an ignition chamber. The high-speed air flows past a tapered tubing that terminates with a fuel-spray jet. Second, the motor drives a pump that forces oil through the tube, expelling it from the jet under pressure. Air and oil-spray mix, forming a heavy flammable vapor that can be ignited in the firebox.

When the furnace is off, you can see the firebox through a little trapdoor in front. The firebox is a large ceramic cylinder, usually molded in two parts. If you look, touch carefully; the access door may be hot if the furnace has been on.

A pair of electrodes in the ignition chamber actually start the oil-vapor mixture burning. A sealed transformer develops a high voltage (very dangerous to touch) and applies it to the electrodes through spring contacts. The transformer usually mounts on the access cover to the ignition chamber.

If you have reason to open the chamber cover, and you may (see next page), be absolutely certain the furnace has been turned off completely at the main electric shutoff. Remove the fuses to the whole circuit if you aren't sure of the switching. You don't want that 10,000 volts alive and waiting to be touched accidentally. The shock won't kill you if you're reasonably healthy, but it could cause you to jerk away and bruise or cut yourself on something else. Avoid the possibility.

The two electrodes, when the thermostat signals your furnace to turn on, create an arc that ignites the mixture of fuel oil and air that is being forced past them by the pump and fan. The ignited mixture is pushed on into the firebox where it burns, making a hot yellow flame. A heat-exchanger shell transfers heat to the air inside a protective shroud. A different motor (not the one that drives the fuel-and-air pump) turns a large blower fan that sends this heated air through the house's furnace ducts.

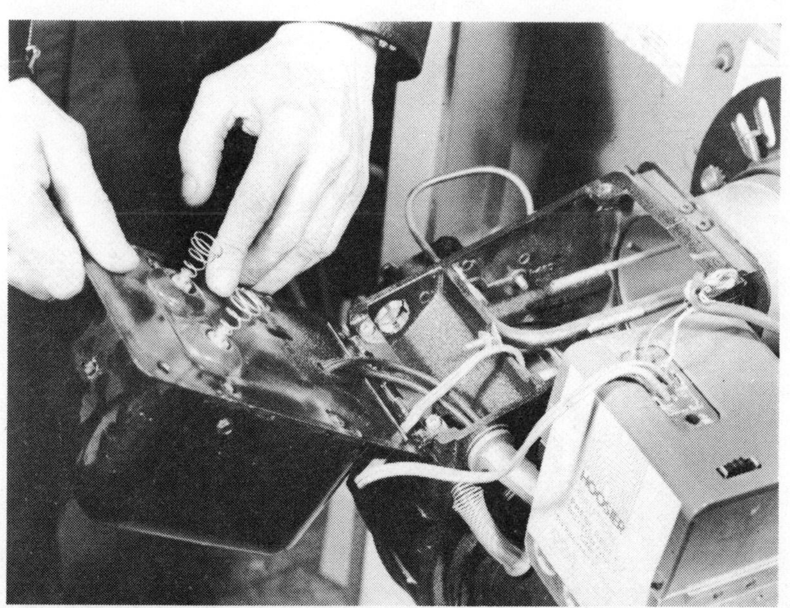

The secret to saving fuel oil—and hence dollars—lies in keeping the furnace in top shape. Even an old furnace can operate economically if you take certain precautions.

First and easiest is the air filter (see page 53). Change it at the start of the heating season and at least every three months. Inspect it monthly and replace if it's dirty. Filters are not as costly as repairs and duct-cleaning.

About once a month, just stop and listen to the furnace operate. Turn the thermostat 'way up for a few minutes. You can hear the burner ignite, but it should be with a smooth roar instead of an explosive or popping sound. The blower shouldn't come on for a minute or two, or even several—whatever time it takes for the inside of the furnace to reach an efficient temperature. Then the blower comes on and hot air begins to pour out of the registers. Turn the thermostat back down. The burner should stop burning quietly. The blower runs several minutes longer. There should be no odor of burned or unburned oil.

With the main power to the furnace shut off, raise the ignition chamber cover and look inside. A formation of soot means burning is inefficient. That and any excessive noises during startup indicate you should call your furnace technician for inspection, cleaning, or repairs.

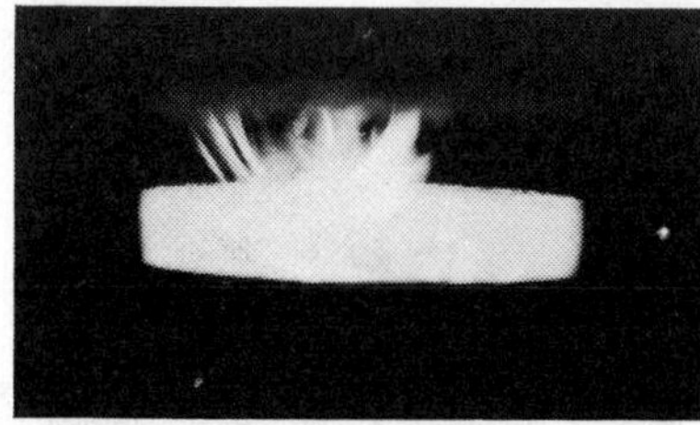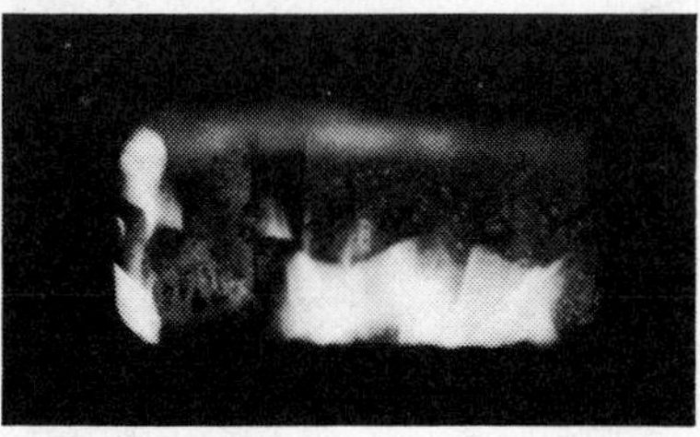

Adjustments that affect efficiency are best left to a technician. Yet, if you use caution and good sense, there are a few things you can check out and take care of yourself. One example is the air-fuel mixture adjustment.

The fuel-pump/air-blower system pulls air for combustion in through a venturi port at the end. An adjustable cover with holes determines just how much air can be drawn in to mix with the fuel spray. All you have to do to adjust the air intake is loosen a couple of screws slightly and rotate the cover to make the holes larger or smaller. Larger lets in more air; smaller, less.

BUT, you have to be familiar with oil-burner flames to make the adjustment correctly. The bottom photos show the flame viewed through the firebox trapdoor. At left, the flame is yellow-orange and lazy, indicating too little air. The flametops may even be sooty, emitting black, stringy smoke. The air venturi cover needs opening up some. The photo at right shows a strong, stiff, active flame, the sign of efficient combustion. The flame is still yellow but without the orangish, lazy look.

If you have doubts about this adjustment, leave it for your technician. Once he has set it during summer tuneup, no one needs to bother with it during the season.

   A fuel-oil odor coming through the heating system signals the need for a technician, usually. The burner jets may be clogged, ignition may be incomplete, or any one of several other maladies may be plaguing the furnace. But get it taken care of. A defective oil furnace can deliver fumes that are harmful and dangerous.

   Still, before you phone the technician, check a couple of things yourself. For example, there's no need to call in an expert if a minor leak is the problem—at least not until you've tried to tighten the joint yourself.

   In the interest of avoiding fuel waste, inspect the fuel lines to the furnace every month or so, even if you don't smell anything unusual. Look the tank over. And make sure your oil deliveryman doesn't overfill the tank (unlikely in today's atmosphere of scarcity and allotments).

Here are a couple more oil-furnace tricks you should know about. Above shows the cutoff knob for the hot-air blower—the one that pushes air throughout the house. The blower shouldn't turn off until most of the effective heat has been blown from the furnace heating compartment or shroud. The control determines the temperature at which the blower turns off. It should be set to turn off the blower when air from the nearest register no longer feels warm. If you're in doubt where to set it, leave it at the center of its range until your furnace technician can do it. In some furnaces, this knob is not out where you can mess with it anyway.

Below you see an overload reset button. Some are round and you push them in to restart the furnace. This one, you push toward the rear. If the furnace has kicked off and setting the thermostat higher doesn't start it, push this button ONLY ONCE —never several times. Each time you push it and get no burner ignition, a charge of oil/air has been sprayed into the firebox. With an overcharge of fuel, if ignition does then suddenly occur, you'd have oversize flame long enough to damage the furnace seriously and possibly even start a house fire.

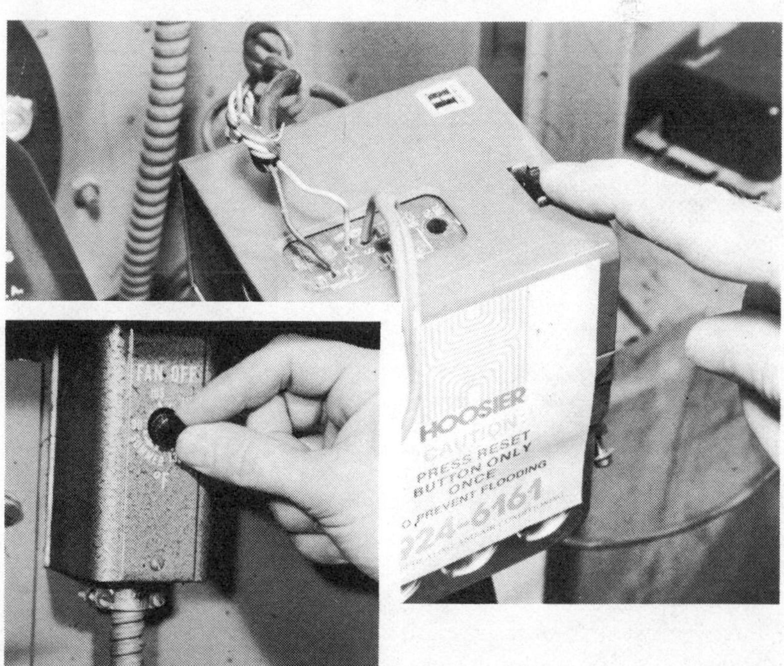

And now . . . about gas furnaces. These have a pilot flame that lights the burners when the thermostat signals the need for more heat.

Gas burners come in two types—for liquified petroleum (lp) gas when you're not near a city gas main, and for natural gas. The difference is slight; they look alike, but you can't use one type of furnace on the other type of gas without technical alteration.

Gas-fired hot-air furnaces resemble oil furnaces in some respects. There's no fuel/air motor, just the large blower for circulating the hot air into the ducts and through the house. Maintenance is simpler, and bits of it you can do yourself. It's not a bad idea to have the system checked by a professional at least every year at the start of the heating season, even if you do some of your own maintenance. This assures you that nothing has been overlooked that might impair efficiency or safety.

There are filters to change (page 53), just as often as with oil furnaces. The dirt they take out of the air has little to do with the kind of furnace; it comes from the environment—dust, industrial smog, and the like.

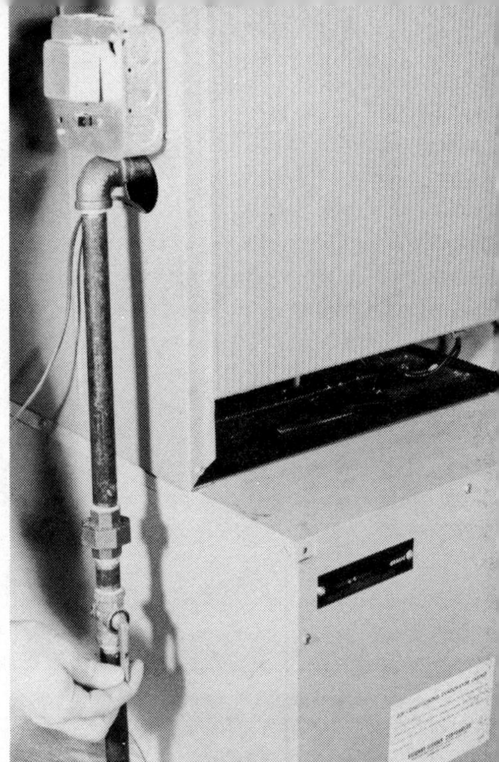

Gas odors are distinctive. Don't attempt repairs yourself if you smell gas. Above all, don't go hunting for leaks with a lighted match. If the leak is larger than pinhole, or has accumulated awhile, you could cause an explosion.

The safest thing to do is shut off the gas at the furnace's main supply valve. That's in the gas pipe, near the furnace. Aligning the handle with the pipe leaves the gas on. OFF is crossways. If there's no separate furnace shutoff, turn off the gas at the main shutoff, even if it means doing without your gas range or hot water for a day or so. Once the gas is off (or even if you can't get it shut off), phone the gas company and your furnace repair technician. Both are concerned about gas leaks.

For added safety, throw the furnace electric switch off too, until the gas leak can be found and repaired.

Beside the switch you'll commonly find a fuse. Anytime the furnace pilot light is okay yet the furnace burners won't fire up, check this fuse. Safety interlocks prevent the burners going on unless electricity is ready to run the blower when the air/blower compartment reaches heating temperature.

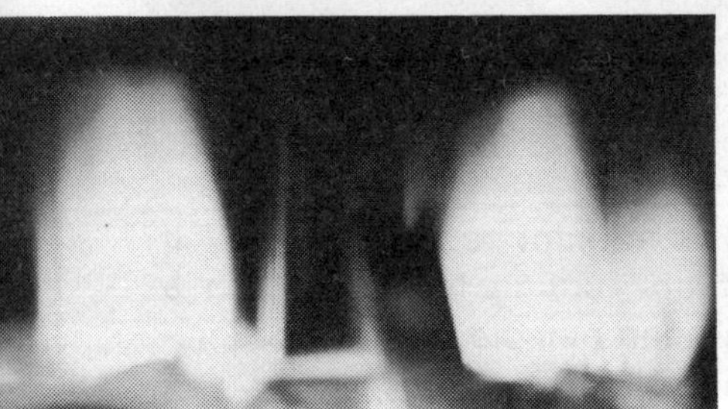

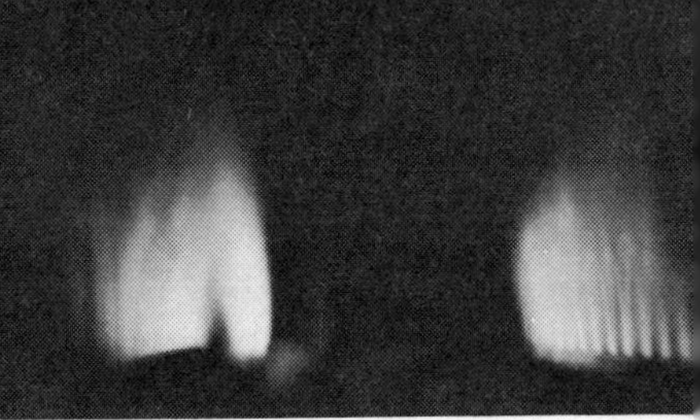

Gas flames need air just as oil flames do. And the gas-air ratio is controlled very much like in oil furnaces. A slider cover over a venturi port for each burner bank lets in more or less air for efficient fuel consumption. An incorrect setting can waste up to 30% of the fuel burnt by the furnace.

The top photo shows the adjustment (after which remember to tighten the screws that hold the sliding covers). The bottom left photo shows the lazy yellow gas flame that signifies insufficient air. At bottom right, the stiff blue flame with just yellow tips tells you enough air is reaching the burners and mixing with the lp or natural gas.

As with oil burners, your technician makes this adjustment when he checks over the entire system each summer. It needn't be altered afterward, sometimes not for years.

Other minor maintenance keeps your furnace operating efficiently for long periods without repairs. For example, a loose
and sloppy V-belt between blower fan and motor can waste the
heat being generated in the furnace by allowing slippage and
thus poor hot-air distribution.

Before you come near a blower fan to check it, turn the furnace's main electric switch OFF. If the blower were to kick on
while your hand is in the wrong place, you could lose a finger
or hand in the ensuing tangle. With the switch off, and the
blower stopped, push inward on the V-belt, about halfway between the motor and the blower pulley. If you can push it more
than an inch, the belt is too loose. Usually, moving the motor in
its mounting tightens the belt. But the belt may be stretched,
and tightening would only wear it out faster. If the belt shows
any signs of age, replace it or have your technician do so.
V-belts are inexpensive. Some blowers are driven directly and
have no fan belt.

At the start of the heating season, and then each month while
you're using the furnace, put two drops of oil in the motor oiling
cups. Each fan bearing may have an oil cup too; three or four
drops in each is a good idea. Some motors and fans can't be
oiled. Talk to your technician about how they should be lubricated.

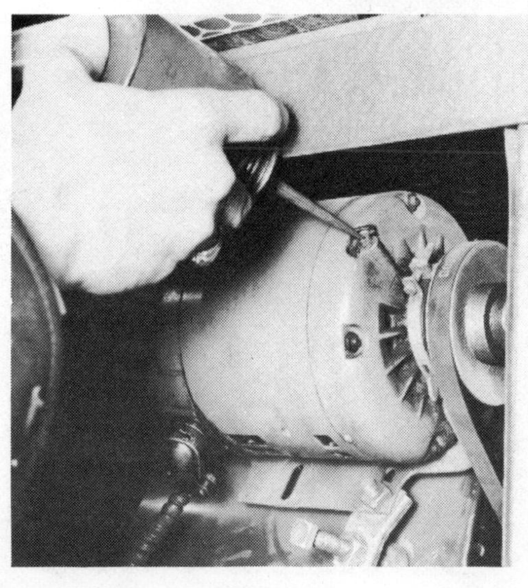

Electric heating comes in various types. There are ceiling radiant systems, a design in which heating elements are imbedded in the ceiling as the house or apartment is built. More common are baseboard heating systems, with heating elements in long banks along the wall near the floor. But the most popular electric heat for homes comes from an electric hot-air furnace. It resembles the others except for having electric heating elements rather than oil or gas. Electricity supplies the "fuel" energy.

You conserve electric-heat energy the same as with other kinds of hot-air systems. Keep the thermostat as low as is comfortable. Keep the filters changed so the blower doesn't have to pull air through a clogged one. Listen sometime as you turn up the thermostat. An audible click occurs as the heating elements turn on. In only a few minutes, the blower should start. And, as with other hot-air systems, the blower runs awhile after the thermostat signals that temperature is sufficient (but there's no click when the heating elements turn off).

If blower fan and motor are separate, although this is uncommon in electric furnaces, check the V-belt. Also drop some oil in the oiling cups, if they're accessible.

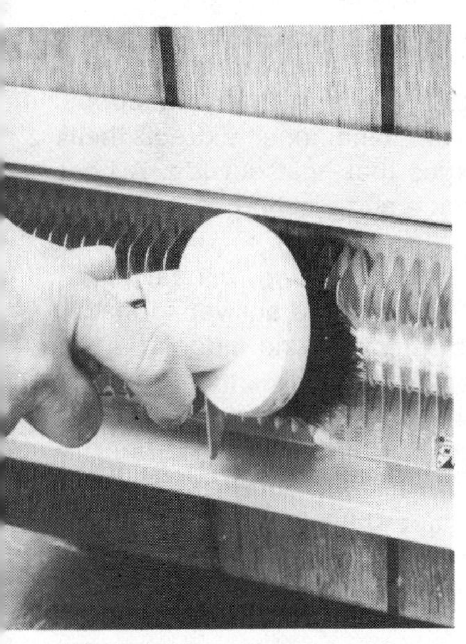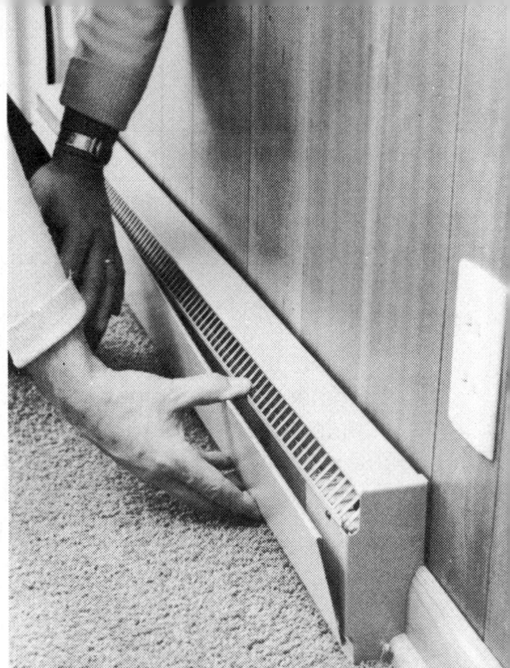

Efficiency of baseboard heating can be seriously impaired, and lots of energy (and dollars) wasted, if the heating elements are allowed to accumulate dirt deposits. It's not a sign of poor housekeeping that these attract dirt. The physics of how they heat just naturally pulls dust that settles anywhere nearby. Convection makes the heated air rise, and cold air rushes in to replace it, bringing along anything light enough to be moved.

The cure is a monthly (or as often as they seem to need it) vacuum-cleaner treatment. The front panel usually just snaps off, or it may be held by screws. The brush-and-hose attachment of the vacuum cleaner does the best job of loosening dirt that has adhered to the fins. (You can use this same cleaning technique with hot-water radiators that have fins like these. Just don't put the brush or your fingers in there while the fins are hot.)

Beyond the maintenance suggestions on these two pages, you had best phone your electrician or electric-furnace technician about any heating or air-conditioning difficulty you have with an electric system. There really isn't much to go wrong, and those few things make a job for an expert. The clue to savings is: don't put repairs off. The energy you waste may cost more than the repair—and the repairs have to be done eventually anyway.

That's not an air conditioner in the photo, it's a device known as a *heat pump.* Looking at it inside, you might still think it's an air conditioner. In fact, it operates similarly to an air conditioner, but in exact reverse. An air conditioner extracts heat from the air indoors and discharges that heat outside. A heat pump takes heat from the air outside and discharges it indoors through a regular hot-air duct system.

When the temperature drops below freezing, a heat pump becomes too inefficient for practical use. The answer to that is supplementary electric heating elements in the indoor part of the system, where the blower can push air over them into the ducts. An advantage, though, is that the heat pump can be reversed in the summertime and used as an air conditioner.

Heat pumps have not become popular yet, but their efficiency has attracted some users. Their major drawback centers around the need for supplemental heating in colder climates, which in wintertime includes most of the United States.

You can do some things with your heat pump to assure economy of energy and heating bills. Air filters are very important to heat-pump heating and cooling; change them monthly, the year around. Fan motors indoors and outdoors must be lubricated, likewise fan bearings. Check V-belts if they're used between motors and blower fans. Outdoors in particular, wipe the fan blade clean of dirt that accumulates.

Outdoors, watch for refrigerant leaks around the piping inside the unit. The refrigerant has a tiny amount of oil in it, and you can see the deposit this oil leaves when a leak occurs, even if it's a very slow leak.

Once in a while, listen to how the unit operates. If the outdoor unit kicks on and off rather frequently—like every few minutes —there are several possible faults: dirt or leaves in coils, frost closing up the outdoor coil (some frost is normal when the temperature is between 20° and 35° outdoors), disabled fan or motor, or insufficient refrigerant (Freon). Some of these you can cure, obviously. Most are for a heat-pump technician (not a regular air conditioner technician).

Heat pumps are relatively trouble-free, but they do require minor monthly attention—and expert diagnosis when the operating pattern varies drastically from normal.

In essence, the way to beat the heating energy shortage lies largely in knowing how your heating system should operate and then making sure it does. You need not be uncomfortable, as long as you know what it takes to maintain comfort.

To iterate: Wintertime room-air temperature should be somewhere around 72° if you're bedridden or merely sitting around reading. With warm socks and sweater, maybe 70° will do. Even slight activity means you can reduce the room temperature to 68° and feel just as comfortable. At night, anything from 60° to 66° can induce pleasant sleeping, depending on how much cover you feel comfortable under—and it doesn't take much if you layer it.

In any case, the game is to be comfortable while saving fuel or electric energy and reducing your heating bills. Your family can be cozy despite the expected shortages. In fact, the very next chapter offers some charming and simple ways to be warmer and more comfortable without wasting any energy at all.

70

Chapter 4

Some Old-Fashioned Ways to Conserve Heat

The energy crunch is bringing back the "good old days." Home "back then" was a symbol of warmth and stability. Yet keeping the old homestead warm was no simple matter. Scarcity was the rule and home heat was conserved by a variety of crafty measures.

Rooms were dominated by massive fireplaces. Around them were broad hearths, romantic tapestries, elegant libraries, and the deep, cozy featherbed. These accouterments were not mere visual pleasantries; they insulated and protected and warmed. This chapter returns you to those days, to help you reconsider the practical ingenuity of our ancestors. They knew how to make-do. You can be that ingenious yourself, conserving energy and saving money as you do.

Houses were drafty. Some of today's are, too. Drafts reduce heating (and cooling) efficiency and ultimately cost someone a great deal of money. One aid back then was window shutters —originally a buffer against wild intruders, but which also cut drafts. You can correct drafts easily if you can find them. The worst spots are around windows and doors, but joints of walls, ceilings, and floors are also suspect. Feel around the seams, joints, and corners in your home. If you feel air of a different temperature, that's a draft.

Almost every hardware store or home center offers a variety of materials to prevent drafts. Instead of shutters (unless you want the decorative kind), attach a piece of clear plastic drop cloth inside each window with weatherproofing tape. If the plastic cover fills up with air and pulls loose, your draft problem is more serious. Check the fit of the window. If it's crooked, somebody must correct it. If the window is merely off its track slightly, a small adjustment will straighten it.

Check caulking, the putty-like substance around the frame. If it's dried out, clean away the old material and shoot in new compound with a caulking gun. Modern caulking has greater resilience and life than older types. A rolled towel along the windowsill helps, too.

You can detect drafts around doors that don't fit snugly. Secure the hinges and repair weatherstripping. Back in those days, old rugs or sacks at the threshold blocked incoming air. You can do the same today with inexpensive throw rugs or rolled-up towels.

Sliding patio doors offer a particular problem. Size and traffic make a plastic drop cloth impractical. Caulking between frames and wall helps cut drafts. So does sealing off with weatherproofing tape. Weatherstrip the center seam. Storm windows and doors are a wise investment. They last longer than all that plastic and weatherproofing tape.

Drafts develop around plumbing fixtures and heating ducts, and often near unused electrical outlets. Caulking or a carpenter can remedy these cold-air spots.

Sod roofs and log walls kept heat in during cold months and out during the hot summer sun. Those fine insulating devices have been replaced by fiberglass, plastic, and other synthetic materials. They come in blankets, bats, loose fill, and tongue-and-groove boards.

Insulation is rated by an R value, which stands for resistance to heat migration. Your house needs 6 inches of insulation (R-19) in the attic and 3 inches (R-9) in the sidewalls. Of course, the more you add—attic, walls, and floor—the more heat you'll retain. You can cut up to 40% from your monthly heating/cooling bill. You may find bare or thin spots in your attic insulation. Add enough overall to bring the level to R-24, or 8 inches. But don't cover the attic vents; they prevent moisture and condensation that could cause dry rot.

Sidewall insulation is best installed during construction. Adding insulation to a finished dwelling is a job for trained technicians with special equipment for blowing the insulation into the sidewalls. You can also insulate the floor. Each type of construction—concrete slab, crawl space, partial or full basement—requires a different insulation method. An insulation supplier can give you instructions.

In the "good old days," extra rooms were sealed off. Folded newspapers closed cracks around windows, and whole sheets covered the glass. With outside drafts blocked, the door to the room was closed and paper stuffed around it too.

Today, unused rooms, guest rooms, or special-interest rooms that are used infrequently can be closed off. Reduce drafts and heat input to a minimum, then seal the door with masking tape. Most cold air flows under doors, so tape the bottom well. Whenever you need the room, masking tape comes off quickly and cleanly. With drafts eliminated, the room warms up quickly when you need it.

Modern houses lose heat through garage, attic, and basement. Throw rugs help there too. Attic access should be sealed the same as a guest room.

One caution: Don't lower temperatures too far in rooms where water pipes are present.

Many of today's decorating trends come from the past. Like the winged chair. It originated in the 18th century, and the winged back protected the occupant from chilly drafts and cradled the warmth from a fireplace. It's as useful today. Highbacked upholstered chairs are also draft-dodgers. Upholstered furniture is not only more comfortable but warmer than plastics. Don't let any furniture block heating vents.

Wall coverings such as tapestries and large pictures help insulate and seal against drafts. Today, carpeting on walls serves the same purpose. Large pieces of furniture, such as hutches, bookcases, tall dressers, and buffet cabinets, placed against outside walls, act as additional insulators. Filled, they insulate even better. Shelves of books are an excellent insulator.

Cloth lambrequins or shadow boxes around windows divert the escape of warm air. Paneling on an outside wall, installed with furring strips, also helps insulate. Furring is 2 × 2-inch strips of wood between the paneling and the wall.

Wall-to-wall carpeting helps warm floors where cold drafts settle. Carpeting is a must for concrete-slab floors, or in rooms for the elderly or infants. Lined drapes close out some chill. Blinds at the windows also help block drafts, and some are decorative. Cafe curtains let the sunshine in while offering privacy and protection from drafts.

Furniture placement rules haven't changed. But today's heating methods have. So you must pay attention to air flow.

For example, cold-air return ducts should never be obstructed—not by furniture, not by carpeting, not by drapes. You may arrange your furniture with good intentions. But somehow, little by little, things accumulate. Eventually, heating efficiency becomes impaired. Without free-flowing air intake, the furnace cannot heat adequately.

You might arrange your furniture to make the duct opening less conspicuous. Place an occasional chair or table with free-standing legs a few feet in front of it. The lack of bulk at floor level will allow free air flow. Heat registers should not be blocked either. Larger pieces of furniture in the middle of the room can foul up the air-flow pattern. Partitions should be placed where they can't interfere with air passage.

Before we wrap up this chapter on old-fashioned ways to conserve heat, it's only fair to give at least a moment to modern ingenuity. The geodesic dome is one sample. Almost anybody can build one. Domes require about one-third the material of conventional structures of equivalent living space. Windows in the top of the dome, on an angle that follows the sun, help heat the interior in winter. Architects suggest that a glass wall on the southern side of houses in colder regions would help warm the place in winter.

Another example of technological and ecological advancement is the IDS Tower in Minneapolis. The photographer has made this beautiful structure appear transparent. Actually, it is constructed with reflective glass. Here are the energy-conserving factors derived in using reflective glass: (1) The reflective finish turns back the direct rays of the sun, trimming air conditioning needs. (2) Because the sun's rays are deflected, furnishings are protected from fading. (3) Reflective film added to existing glass not only reflects heat but also makes the glass shatterproof.

Reflective glass has only begun to be used for residential building. Cost is the factor. Reflective glass is similar to one-way mirrored glass. You can see out, but passersby cannot see in. The glass thus offers its own privacy during daylight. But if you forget to close drapes at night, you are in a goldfish bowl. A timer set to close the drapes at dusk takes care of that easily.

Reflective glass comes in varieties: insulated double-pane glass such as Thermopane, single-pane glass for warmer climates, and an easy-to-install film called *solar film.* You can install solar film yourself. Cut it to the proper size with household scissors. A spray-on adhesive has the color and consistency of water. The film then goes up to the glass, where you smooth it with a squeegee. It's that simple. The heating and cooling savings will more than pay for the film in one season.

Staying alert to product innovations like these enables you to conserve as much of our vital fuel resources as by drawing on ideas from the past.

The open flame, the oldest method of home heating, is still alive. Here, to end the chapter, are a few fireplace do's and dont's.

Always use a wire mesh or glass fire screen. It will catch popping embers.

Use only wood that has been properly dried. It is less likely to spew embers.

Clean soot from the chimney at least once a year. Or, occasionally shake a bit of copper sulfate onto the logs. The pretty blue flame releases fumes that keep the chimney free of soot buildup.

Never burn charcoal in a fireplace. It gives off carbon monoxide.

Never use gasoline, charcoal lighter, or other highly flammable fuel to start a fire.

Don't burn trash, loosely wadded newspaper, or cardboard boxes in your fireplace. Paper can be burned safely if twisted very tightly into logs. Do not soak paper logs with fuel oil; this is highly dangerous.

Keep furniture and rugs back from the fireplace. The heat from a properly maintained fire can be felt for several feet.

Do not stack firewood beside the fireplace. It can be ignited by sparks. Also, insects in the wood may take up permanent residence in your carpets and walls.

Never leave a burning fire unattended, even with a fire screen. Start your fire early enough that it will be just embers by bedtime. Turn the damper down, but not off, to keep the embers controlled as they burn out.

A nice bed of ashes makes the best fire, but keep your fireplace clean enough that downdrafts don't blow ashes back into the living area. Keep the ashes about an inch from the bottom of the grate.

If you don't use your fireplace, close the damper. Seal off the fireplace with a piece of Masonite or plywood. You can paint the board and decorate with potted plants.

Chapter 5

Your Own Energy Can Save You Money

One energy source is still relatively plentiful and inexpensive. It's your own personal energy. Properly utilized, your personal energy can bring you more money, more relaxation, and more time for yourself. The keys are good health and vitality, plus intelligent scheduling of your work and play times.

Shortages open new opportunities to use this vast resource—this limitless energy in ourselves. Take advantage of the energy "crisis" to improve your way of life. Learn about clothes designed for real comfort. Learn the art of scheduling your everyday activities better. Enjoy life more fully. Some pointers are right here in this chapter.

Your resilience to new temperatures can be developed through a routine of daily exercise. Just make sure your physician says it's okay. He may suggest special exercises especially for you.

Some suggestions: Breathe deeply three times before you start. Exhale fully each time. Stretch to the ceiling, to your toes. Emulate a cat stretching after a nap.

Try some deep knee bends with your arms straight out. Turn right and left to get your waist into action. Place your hands together over your head and bend to the left and then to the right. Go further as your body limbers up.

The upside-down bicycle exercise is excellent for general circulation, hips, thighs, and complexion.

Start with about six times for each exercise. Vary the routine every couple of weeks, to work different muscles. You probably will lose pounds and rearrange some inches. That's a bonus.

Daily exercises are no panacea. They merely help. If you feel better, you'll look better and work more effectively.

If you lack the persistence to stick to a daily routine, get with a group of your neighbors, or go to the gym or "Y" on your lunch-hour.

Your clothing also helps you adapt to new temperatures. Layered clothing for colder weather is sensible, comfortable, and fashionable. Try cotton shirts and blouses covered by vests, overshirts, or sleeveless or cardigan sweaters. Cold air on the legs calls for slacks, knee socks, maxiskirts, or a mixture of midilength skirts and boots. Pantyhose help, whether with slacks or skirt. Three-piece suits are good for men, women, or children. Choose layers of clothing that do not restrict body movement. Evenings, when you're less active, afghans, house robes, smoking jackets, and flannel nightwear all help retain body warmth.

In the really colder months, thermal underwear makes sense. Women wearing slacks and long-sleeved blouses can enjoy this same comfort. Outdoors, insulated hunting clothes are good, too. If they're not your style, wear warm coats, boots, stockings,

scarves. Don't wear really long scarves; they can get caught in precarious places. Ski masks are good for extremely cold days. Lined parkas and hooded coats utilize body heat better than coats and hats. You need something on your head because a great deal of body heat can be lost there.

Hats are also good for summertime, to keep hot sun rays from your scalp; wear straw types, with wide brims. Natural fibers such as cotton and light wool or linen act as a blotter for perspiration. Breezes evaporate this moisture, which is very cooling. White and light-colored clothes reflect heat and so are more comfortable in the summer. Save dark colors for cooler weather. Sandals are more comfortable in summer than enclosed shoes.

Scheduling your days serves two energy-conservation purposes. For one, it's better to get housework chores done before the hours of heavy demand on the electrical supply—about 11 a.m. The other has to do with your personal warmth. Getting busy early in the morning warms your body up quickly. Also, the heat from appliances helps take overnight chill off the house. You'll find you can keep the thermostat at the low nighttime setting until after your morning activities, if you dress warm when first you get out of bed. Save your morning bath for later in the day when the rooms are warmer.

Monday morning is a good time to clean up after the weekend. The house is more likely to stay clean longer than if you try it on Sunday. Friday, clean for the weekend.

Schedule laundry so you can wash and iron consecutively. During the first load, vacuum the floors and rinse the breakfast dishes for storing in the dishwasher. The laundry won't leave enough hot water for the dishwasher; save them till evening. When the first load of laundry is damp-dry, iron.

By getting chores done in the morning, you leave the afternoon for shopping, card parties, sewing, or tennis. From 11 a.m. to 5 p.m. is the peak period of electrical demand for business. During this time, choose activities that require personal energy, not electrical energy.

In the evening, after dinner is cleaned up, take advantage of natural body temperature as well as room comfort. Do your reading and quiet activities early in the evening, while daytime and kitchen heat still have the room warm. Save that rousing game of ping-pong till later when the rooms are cooling off to nighttime temperature. Do fix-it chores later in the evening; getting them done early wastes the warmth generated by your work.

This kind of thinking may seem unusual for a while, but it can make your entire life during the energy shortages more comfortable. It's only a matter of rescheduling old habits.

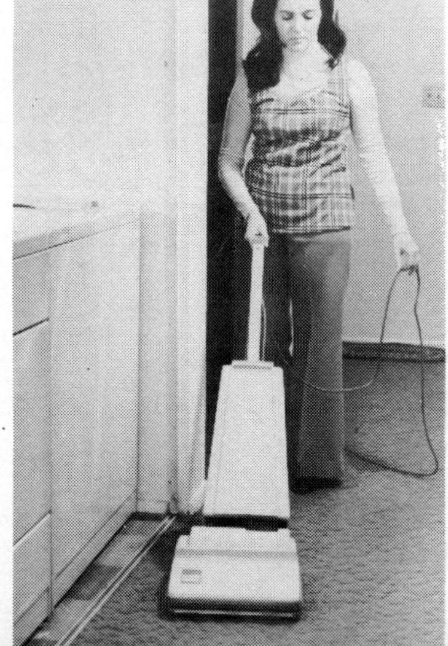

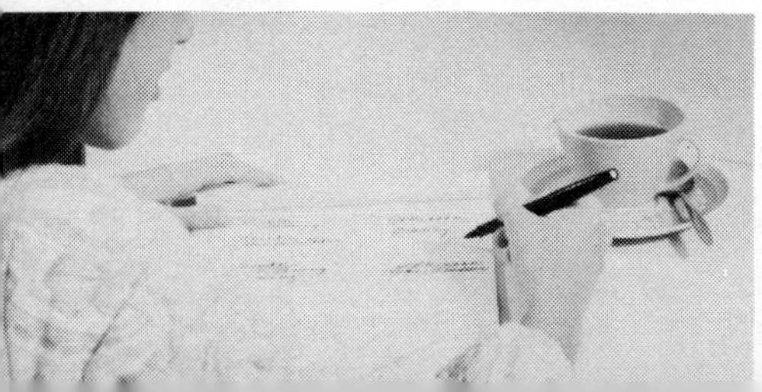

Since you turn down your thermostat when you go to bed, you can save more heating fuel if you go to bed earlier. You can catch up on your reading there, or watch television, with the blankets and a cozy robe to keep you warm. It's an excellent time to study. There's a rumor that your subconscious works on whatever you read or study just before going to sleep, implanting it in your mind.

If you do go to sleep earlier, you might wake up earlier. This is a particularly good idea for summer. You can finish the hot chores early in the morning before the heat of the day begins. Your air conditioning unit won't need to work as hard.

There are many ways to build up and conserve your personal energy. Some are very subtle. For example, consider commuters and others who travel a lot.

One hour of travel to work is common. That's two hours a day, ten hours a week, more than 40 hours a month—an entire work week. You could keep up with your reading. One industrious woman wrote a book during three years of commuting—certainly a wise and profitable investment of her time.

Take stock of your routine. See where precious time may be escaping you in other necessary "nonactivities." You can put your personal energy to work either making more money or otherwise improving your lifestyle. It works. In the process, you'll find coping with shortages and high prices much easier. This one, you owe yourself.

Chapter 6

Combating Food Shortages and Prices

Food scarcity began with beef and pork in 1973. Other products followed. We may have to learn to live with uncertainty in the food supply and price situation. One packers-union official says, "There is no need for us to have shortages. If a year is good weatherwise, and the government will allow supply and demand to function fully, supplies of all goods, including meat, should be ample." Nonetheless, statistics from the U.S. Dept. of Agriculture show fewer hogs and cattle on feed than in prior years.

Whatever the excuses, we have high prices on almost every food item we purchase. Whatever the scarcities, or the predictions, or how long they last, we have to live with them, and manage on our given incomes.

As you've seen, to save anything—money, time, or energy—you must have a plan. Forget impulsive shopping at the supermarket. A vague idea of what you need won't do. It may sound old-hat to check newspaper ads, plan menus, make a list . . . but do it if you are serious about stretching your food dollars.

Watch out for those "instant" items. You pay for convenience, and you have to weigh energy (cooking) savings against dollars. The following pages should help you make meal preparation smoother, quicker, and more economical.

Skillful use of substitutes can stretch your food budget. Of course you may already be into the new protein substitutes (page 106), or the natural protein extenders for meat. Additives go well with ground beef. You can purchase them in packages or rolls, sometimes in gallon cartons. The beef additive may be frozen or flaked, and broken off as needed. Goes great in chili, spaghetti sauce, patties, casseroles, or stews.

Remember noodles, rice, and spaghetti can be mixed with cheeses. Makes a good meat substitute at any meal. Longtime favorite macaroni-and-cheese is still relatively inexpensive, and a fine protein source.

Don't forget the once-lowly bean and its cousins: lentils, peas, blackeye peas. Likewise the versatile potato. Bean prices have shot up wildly, but a few pounds of beans still go a long way as an alternative to costly meat protein.

Meat isn't everything. When was the last time you fixed a big stewpot of good old, plain old turnips? That pretty purple-and-white vegetable has a new taste cooked next to a bit of pork.

Tried cauliflower lately, with or without cheese sauce? How about cooked celery served in a white sauce? Fresh carrots peeled, sliced into brown-sugar syrup, and served glazed? Don't forget parsnips, artichokes, asparagus, broccoli, spinach. The entire gleaming, ice-cold vegetable department awaits your selection.

Choose nutrition, and treat it glamorously. For example, peel a large eggplant and slice it thin. Salt the slices and stack them together for 30 minutes. This eliminates the bitter tang. Drain off the liquid and cut the largest slices into quarters. Peel and slice six large tomatoes and five medium onions. Arrange a layer of onion slices in the bottom of a large baking dish. Then a layer of eggplant. Then a layer of tomatoes. Season each layer with salt and pepper. Fill the dish with olive oil to the top layer of tomatoes. Bake three hours at 300°. Sprinkle with parsley and serve hot.

Have a closed crisper jar or tray, and keep it full of vegetable sticks . . . always chilled and ready for nutritious, nonfattening nibbling. Celery in the crisper is always iceberg-ready to be filled with cream cheese or peanut butter.

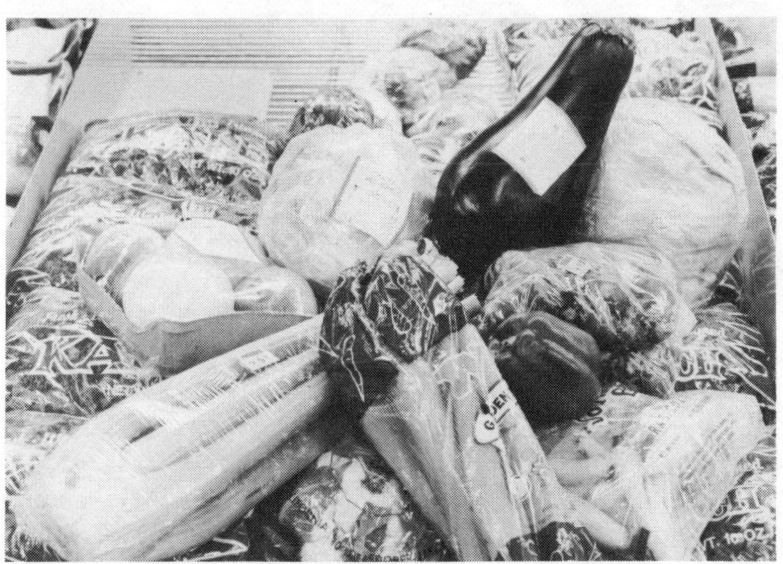

Hamburger can easily qualify as king of the save-money meat budget. Say you bought 3 lb of hamburger. What do you see? A mound of meat to make patties for one meal, or the makings of a 3-lb, two-meal meatloaf? Try seeing it as 1½ lb of meatballs and ¾ lb made into delicious hamburger gravy to pour over biscuits, rice, noodles, or baked potatoes.

Still ¾ lb left. Brown this. Add onion, diced green pepper, and chili sauce, and you've got Spanish rice. That's three meals, possibly four, from one average mound of hamburger.

That's not to mention the casserole hiding in the leftovers from those meals. Like so: Take what's left of the meatballs, layer it in a casserole, and freeze it. After you serve the Spanish rice, add a layer of that leftover. Do the same with the leftover gravy, noodles, whatever. Then, one busy evening, slide this bonus casserole into the oven. Fix a green or yellow vegetable, or both; add cottage cheese, lettuce wedges, and garlic bread. An almost-free meal.

If you're flexible enough, try tuna with a hamburger-helper dinner. You can juggle all sorts of bits of meats, canned or leftover. Dice up an onion, green pepper, celery, or the like to stretch the whole deal. They add the extra touch of flavor that makes a meal your individual creation.

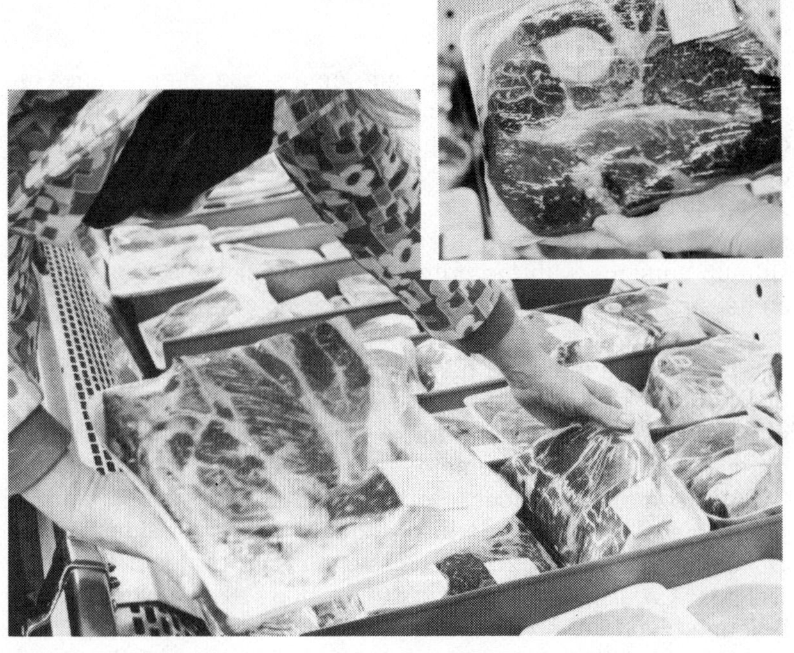

You can adapt your creative ways to beef roasts. Choose an English or Boston cut chuck roast, or a boneless shoulder pot roast or steak. Watch the weight, how much bone, and the price. You may prefer whichever cut is on sale.

Make your shoulder roast exotic with sprinkles of clove, dashes of soy sauce, onion slices, a quarter-cup of orange juice, and two cups of black coffee. Keep it in a 275° oven all afternoon, spooning liquid over it occasionally. Add water as needed.

For a more colonial dish, brown the meat lightly, arrange sliced onion, diced green pepper, and diced celery around it. Cover with one can of bisque-of-tomato soup, one can of golden mushroom soup, and two cans of water. Simmer nearly forever at 275°.

Either of these dinners could have quartered potatoes baking with them, coming out literally smothered in sauces. Or . . . peel a potato for each serving. Leave them whole, but score them across the top. Pour a spoon of melted butter over each one. Sprinkle with a mixture of salt, pepper, celery salt, and paprika. Pour about an inch of water in the bottom of a casserole dish. Cover, and bake until tender.

By watching for meat specials, and knowing how to stretch them, daring to combine and really create, you can afford beef.

If you are a bacon-and-egg buff, or a sausage-and-pancake connoisseur, eating on the hog has become mighty high indeed. The USDA says little-pig will soon get to market more often, and that should help level prices off.

Meanwhile, jowl bacon costs less than regular. Slice it thin and fry it; crumble it into scrambled eggs, or top your sunny-side-up versions with the crumbs. Study your bacon packages; some show tiny strips of lean and it goes no deeper. Quit buying those brands.

When you have baked ham, freeze the bone. Add it to soup another day or vegetable stew. Simmer it enough and all the pieces fall off. From them, make a ham dressing to stuff into green peppers; pour tomato sauce over and bake slowly.

Save some ham slices. Sprinkle with brown sugar, add a pineapple slice or mandarin orange: a bright breakfast item. Or, cube some of the ham for creamed ham on toast squares. Or, mix a *real* ham salad; most everybody else makes it with bologna, weiners, or luncheon meats. Or, grind the ham up to make ham patties. Include bread crumbs or cereal flakes, and some liquid like milk or orange juice—just enough to moisten and shape. Imaginative planning by an enterprising cook should let one ham take care of three or four meals.

The barbecued pork rib continues to be a taste treat. And how about country backbones? Have you ever deviled a pork chop? Buy end-cut chops or pork steaks, flour lightly, and brown in a heavy skillet. Cover with onions, a can of undiluted tomato soup or tomatoes or puree, or a half-bottle of catsup. Add a little water and bake a couple hours.

Check labels at the meat counter. Sometimes a boneless ham costs little more but gives you much more meat. Same with pork roasts. Boneless rolled roast and Boston butt roast offer meal magic.

Your butcher can divide the roast for you. You might want a few slices off the end to bake with apple wedges, to devil, or to just pan-fry. Bake the rest. Use bits to make a barbecue sauce for hot sandwiches. If there is a bone, save it to simmer.

Treat pork-flavored vegetable protein products as you would already-cooked meat. In a ham loaf, you should use more eggs, crumbs, and milk. Then you bake only long enough to cook the other ingredients, usually a half-hour. Make ham salad with ham-flavored cubes. Or stir the cubes into cheese sauce with olives and green peas.

Ever longed to roast a pig, serve it elegantly with fanfare and an apple? Would you settle for mock piggie patties? Made with ground beef and ground bologna? Add some milk, egg, and crumbs. Mix them well and shape. While they are frying, slice some apples, and place them around the patties. Add a smidgeon of water, and a handful of little red hot cinnamon candies. Soon you have your pig with apple. Great for breakfast, with hot chocolate, a boiled egg, tomato juice.

Beef and pork prices drove shoppers to poultry. Soon, prices for poultry began to rise too. The USDA says chickens, eggs, and turkeys all are on the increase, and prices will lower and stabilize.

Now, even though prices for poultry are up, they are competitive with those of beef and pork. And fowl are an excellent source of protein.

There's an abundance of menu possibilities. Occasionally whole fryers are a bargain. Roast or bake it with stuffing. Or cut it up and fry it. Or just simply boil it and use the meat for sandwiches, salads, or omelets. The whole fryer is a great way to get food mileage from a chicken.

Cubed pieces of chicken can go in a tossed salad, show up in a skillet dinner, and flavor a spicy white sauce in patty shells, The pieces mix well with chicken additive, or the additive can be used by itself like any other cooked meat. Simmer additive cubes in bouillon water, or to use for fricasee chicken.

How come you think of turkey just at Thanksgiving? He's a gold mine of ideas anytime. Take a 12-lb bird and roast to standard golden brown. Do whatever you like about stuffing, seasoning, and all that.

The first meal, you serve beautiful slices of white and dark meat with harvestime favorites: sweet potatoes, cranberry salad, a zesty dressing, green beans almondine.

The next meal might mean chipped turkey in an omelet. Brown some chopped onions in an electric skillet. Stir into them a blend of eggs, milk, and a quarter-cup of mayonnaise. Add turkey pieces just in time to heat through. Preface with ice-cold tomato juice, serve with hot muffins, and follow with a fruit cup.

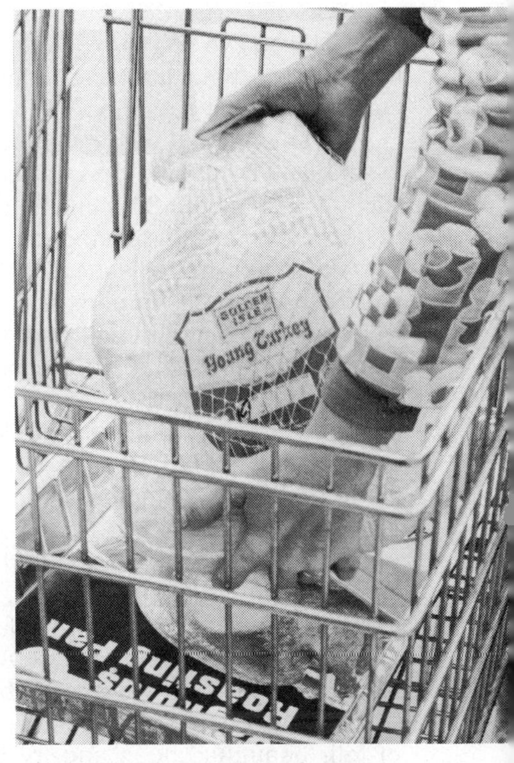

Now ready the bird for his additional performances. Pick as much meat off the carcass as you can. Wrap the bits for another meal—maybe for sandwiches to eat around the fireplace, with a cup of hot cranberry juice. Marshmallows for dessert.

Wrap the carcass and maneuver it into your freezer. One day soon, unwrap and simmer it gently for broth. There's still quite a lot of meat on those bones. You can make a big bowl of dressing. Stir up a gravy from the broth, to serve with the dressing, or with this meal have mashed potatoes.

And, of course, remember to scrape the roaster pan of drippings and turkey bits. This makes a deep dark-brown, rich gravy—the perfect topping for baked potatoes or mounds of fluffy rice.

Turkey salad, turkey in box dinners calling for chicken or tuna, turkey in tetrazzini. You name it, you can probably make it with leftover turkey.

GOURMET TOUCHES ON A SHOESTRING should be the title of these two pages. You can eat well despite shortages and high prices. It's a matter of slight additions to a dish or table setting to enrich mealtime.

Stuff an eggplant with mushrooms. Halve an eggplant lengthwise. Place it cut-side-down in one tablespoon of vegetable oil. Bake at 375° about 30 minutes, until the eggplant is soft. Cool. Scoop out the pulp, leaving the shell intact. Set aside.

Make a rich butter sauce. Melt one tablespoon of butter or margarine in a heavy pan. Add a tablespoon of flour and mix well. Put in a pinch of salt and a half-cup of milk, stirring constantly until the sauce is thick and smooth.

Then mix the onions, mushrooms, and two tablespoons of chopped parsley with the eggplant pulp. Pour in just enough butter sauce to bind. Season with salt and pepper. Bake at 425° for ten or fifteen minutes or until the eggplant is bubbly, hot, and brown.

This will serve four to six, perhaps along with baked country ham or a stuffed baked chicken. It could be the main dish of a meatless meal, served with a colorful tossed salad, hot buttered and very garlicky french bread. Another elegant way to stuff the eggplant is with shrimp and crab.

Give pork chops a Hawaiian vacation. Trim fat from the edges of four beautiful chops and fry in a heavy skillet. Remove the pieces. Flour the chops and brown them in the hot fat. For the Islander sauce, mix a half-cup of vinegar, a half-cup of catsup, a 9-oz can of crushed pineapple, one tablespoon of soy sauce, one tablespoon of brown sugar, and one half-teaspoon of salt. Spoon the fat from skillet and pour the catsup mixture over and

around the chops. Bake at 325° for 1½ hours or until tender. Spoon the sauce over the chops once or twice. Serve chops on rice mounds, and add a little extra sauce on top.

A variation might be Chops Mardi Gras, using canned tomatoes, onion, a bayleaf, three cloves, and Worcestershire sauce. Or bake with a dash of curry in apple cider.

There are many frozen, boxed, and canned oriental delights. But if you're really in the mood to cook, try sukiyaki from the cutting board up.

Have a genial butcher cut two pounds of sirloin into thin half-inch strips, against the grain. Chop twenty green onions. Slice two large green peppers into strips, and slice one large, white sweet onion. Cut up a can of water chestnuts, and dice a half-pound of mushrooms. Set all the vegetables aside.

Prepare two cups of instant bouillon and add to a half-cup of soy sauce, one quarter-cup of A-1 steak sauce, three tablespoons of sugar, and salt and pepper to taste. Open and drain two cans of bamboo shoots.

About twenty minutes before serving time melt some butter in a large heavy skillet. Quickly brown the meat strips in the melted butter, turning frequently. "Brown" means barely let the strips lose their pinkness. Immediately add the bouillon mixture, and simmer gently and briefly. Then add the vegetables, bamboo shoots, and mushrooms. Simmer this for fifteen minutes, or until the greens are barely tender when pierced. If desired, a thickening made from two tablespoons of cornstarch and a quarter-cup of cold water may be stirred gently into the simmering mass to give the juices a light gravy consistency. Chilled tomato juice, Waldorf salad, hot poppy-seed buns, and candlelight go well with this meal.

A late-show buffet or Sunday-morning brunch can feature Buried Treasure pancakes. Use your regular pancake recipe or the boxed mix in your pantry. Retrieve the three hotdogs you've been saving, or dice a can of luncheon meat. Slice the hotdogs like pennies. Add them to the batter, griddle as usual, serve up with homemade brown-sugar syrup. Add a thawed package of strawberries and a small can of crushed pineapple to a large can of fruit cocktail. Presto: a chilled fruit cup.

A salad can be as simple as lettuce wedges with a small bowl of mayonnaise alongside. But gourmet entrees call for something more fanciful. It helps to keep an abundant supply of greens and the other makings: red cabbage, iceberg, bibb, and leaf lettuce, green peppers, the onion family. Don't bypass cooked vegetables in your salad creations: artichoke hearts plain or marinated, green peas or beans, sliced cooked carrots marinated. Consider fish and meats: ham, tongue, cold cuts, turkey, chicken. Cut meat into julienne strips. Slice cheeses into strips or cubes. Think of apples in wedges or slices, sliced avocado, fresh pineapple, banana slices, orange sections, raisins, dried apricots, prunes, nuts. Your imagination is the only limit. Try for new tastes.

Making your own dressing can offer a taste challenge too. Start with a quart of salad dressing, mayonnaise type. Mix in two cups of low-fat or powdered milk. Add one quarter-cup of cider vinegar, some pepper (freshly ground if possible), one half-cup of sugar, a bit of paprika. Mix thoroughly, or blend in a blender.

That's the base to which you can add a jillion things: crumbled bleu cheese, poppy seeds, celery seeds, dill weed, or catsup. Try fruit juice or puree for a sweeter kind of dressing. Or prepared mustard for more tartness. Add chopped onions and pickle relish to go with fish.

Make your own croutons too, either garlic or celery salted. Trim crusts from four slices of white bread. Rye or wheat bread you've been saving in the freezer is okay. Butter both sides of the bread generously; sprinkle with garlic powder or celery salt. Cut into half-inch cubes, place in a baking pan at 425° till crisp and golden.

Some really fabulous desserts appear when you look at leftovers a new way. A few cake crumbs, ice cream, and bits of fruit can turn into a quick crowd-pleaser.

In a cake pan, layer alternate pieces of whatever cake you have. Cover with a layer of ice cream. Sprinkle in some chocolate crumbs. A handful of coconut and some chopped cherries garnish the top.

The point is, don't overlook anything in the refrigerator or freezer as a possibility for a potpourri dessert of this kind. Any ice cream flavor or color works. A smidgeon of applesauce, cookies or donuts to crumble, some red hots to sprinkle on top.

Yellow cake with malt frosting, nut ice cream, half a cup of cranberry sauce, top with homemade caramel sauce. Call it Crumbs New Orleans. Put a cherry on top, and a cup of hot coffee alongside. Turn on a candle.

By now, can you foresee a day you might have some lime sherbet and vanilla ice cream and drop-in guests? Layer the ices in a pan like a checkerboard and cover with chocolate sauce. Or dip small balls of each flavor into custard cups and pour the sauce over. Gourmet desserts on a tight budget.

You have learned the culminating wisdom of shortage cookery when you can look into the refrigerator and not pine after what you're out of. Gems await in bits and dabs of leftovers. Don't save food until it's dried up and useless. There is money to be saved in leftovers. Good eating is a bonus, not to mention creative satisfaction.

The leftover game is based on the principle of using what's available, no matter how wild the combination. Most things can be combined.

One delicious casserole (opposite page) was born from the doubtful union of small portions of peas and corn, ground beef, leftover chicken, and rice. It was arranged in layers beginning with the rice, then the meats and vegetables, and topped with a cup of catsup and some water. A thoughtful addition was onion slices and green pepper rings. A mixture consisting of chili powder, three tablespoons of flour, one teaspoon of salt, and a teaspoon of pepper was sprinkled between the layers of rice and meat and vegetables. It got dubbed Chicken Mexicali. Baked at 350° until rich and bubbling, it was served with a batch of homemade dill biscuits, coleslaw, pineapple slices, and cookies.

This combination probably would never occur again. But you just might find some leftover cooked carrots, a bowl of twenty-five beans, three or four hotdogs. Layer it all with rice, or cooked noodles, or onion slices, or whatever other favorites you keep in the crisper. The potential is endless, and usually delectable.

You might discover something so especially good that you would incorporate it into your planning to assure the same combination again sometime. That's a *plannedover*. Any cook with a casserole up the culinary sleeve holds the cooking ace. Deal yourself another hand anytime.

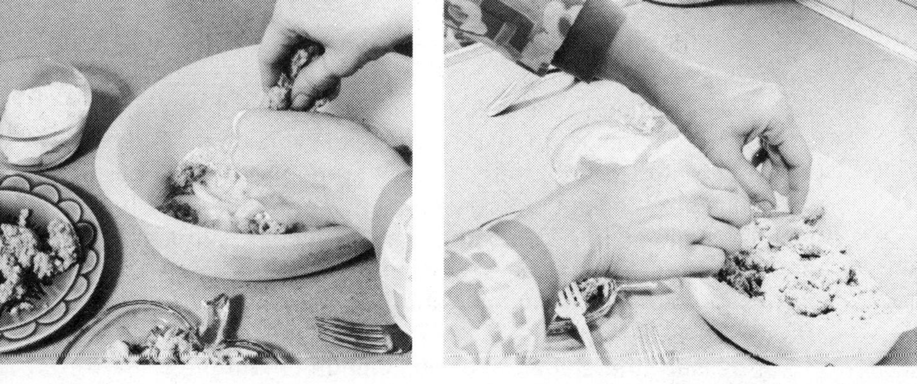

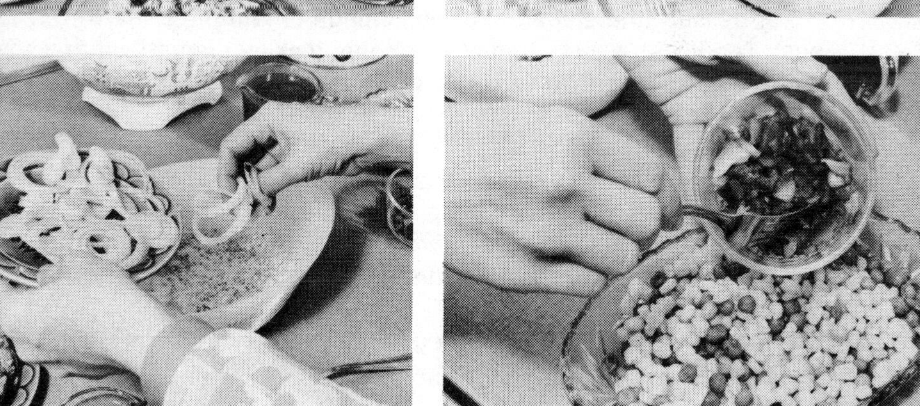

You've seen how to put together gourmet delights from entree to dessert without suffering from shortages in your supermarket or in your budget. It shouldn't surprise you to learn that you can throw a pretty good party for just pennies. There's a knack in making nearly any occasion partytime. There are two rules. Keep it simple but special, and serve it with a smile because you're enjoying yourself.

An informal buffet party needs a sandwich. Let yours be ham salad, but make it with weiners or bologna. Three pounds of either one, ground up with five hard-cooked eggs and mixed with three-fourths cup of pickle relish, one teaspoon of mustard, two tablespoons of sugar, and enough mayonnaise to blend well, makes enough spread for twenty people (with seconds for a few). Either make up the sandwiches ahead, or serve the bowl and let those who prefer use the spread as a salad garnish or a chip dip. Set out a large bowl of corn chips, and have a backup bag in hiding. You can make a zesty dip from a package of cream cheese, a carton of sour cream, a cup of mayonnaise, dashes of paprika, garlic salt, and poppy seeds. Offer a relish tray if you like.

For dessert, the elegance of homemade cream pie is hard to beat. Top it with a mountain of meringue; makes it mouth-watering, regal, and hard to turn down. Butterscotch is a perennial favorite. Crusts can be made ahead, or the frozen-food counter carries them at reasonable prices.

Here's how you make an exquisite filling: Mix one cup of brown sugar, a walnut-size lump of butter, and one tablespoon of cream. Cook until caramel. Gradually stir in 1½ cup of milk, two tablespoons of cornstarch, and two beaten egg yolks. Cook until thick, stirring constantly.

For meringue, whip the two egg whites until they make stiff peaks. Pour the cooled filling into the baked shells. Top with the meringue, and quick-toast at 425° for 3–5 minutes.

Most cookbooks have recipes for vanilla and chocolate puddings if you want to make these flavors from scratch. However, boxed puddings give good results. Mix banana and vanilla and you have a banana delight, with or without the real fruit.

Plenty of hot coffee, a punch bowl, and mixed drinks for the adults, should round out your party. Quick punch consists of two packages of Kool-Aid, a can of fruit punch, and a large bottle of ginger ale. Coloring in the punch carries the party theme: red punch for Valentine's Day, green for St. Patrick's, etc. Placemats could be made from red or green construction paper. Let the kids cut out hearts, shamrocks, cupids, leprechauns, or whatever. Keylime pie goes with St. Patrick's Day, even to green meringue. Pink meringue on vanilla cream pie for Valentine's Day. Or pink cheesecake.

In October, with homecoming games for highschoolers, substitute the traditional sloppy-joe for ham salad. But make it with the soya extender, with or without a bit of ground beef. Make a pumpkin spicecake. Frost it with orange-colored butter-cream icing. Centerpieces and table decor ideas can be gathered up from around the house. Let your imagination run rampant. It's partytime!

Ever since the Earl of Sandwich or somebody put chunks of meat between two rolls, the "sandwich" has held universal appeal. Everyone has his favorite. Occasionally, as with the hamburger, it becomes a favorite everywhere. They come in all sizes and shapes: open-face, double-deckers, heroes or submarines, supers. From barbecue to peanut butter to steak, you can have almost anything you like on a sandwich.

Many a fantastic sandwich lurks unsuspected in your refrigerator. All they need to get them going is an appetite and a creative mind. Let's call some of the more exotic ones *sandwiches supreme.* Like any other menu, they can be created from leftovers, from scraps and bits, from plannedovers. You'll find enough to make into quick hot or cold sandwich meals, anytime, day or night.

Layer crisp bacon slices, large but very thin slices of sweet onion, and a thick slice of swiss cheese onto an English muffin spread with horseradish mustard. Add pickles, corn chips, and beverage: sandwich supreme!

You might want lettuce too, and a bit of catsup on that bacon creation. If your bacon supply is limited, stretch it by crumbling it into pieces and mixing them with some mayonnaise or peanut butter. The nutty flavor makes an unusual change.

Make an instant Hawaiian sandwich. Fry slices of luncheon meat, canned or cold-cut, along with pineapple slices. Pour on a light covering of barbecue sauce. Maybe you would prefer the meats diced, and maybe the pineapple too. How about diced onion simmered in barbecue sauce? Serve the result on toasted buns, corn muffins, or whatever you have around. This is a great way to use up a couple slices or chunks of any kind of leftover meat.

You can concoct another elegant hot sandwich from tuna folded into hot cream-of-shrimp soup. Serve over slices of cheese and onion on two halves of English muffins. Put some kind of sauce on top. Add some pimiento-stuffed olives, sliced or minced. A brunch favorite, or a perk-up for lagging breakfast appetites, can result from using crabmeat, or leftover chicken or turkey, and perhaps other kinds of cream soups.

This takeoff on a nationally sold breakfast delight starts with a hot griddle. Start a nice fat egg frying alongside a slice of bologna. Canadian bacon or ham is better if you have it. Lay two slices of bread on the same griddle. Just before your egg finishes, lay a slice of cheese on the bread to melt. Stack the whole thing together and serve with your favorite preserves or jelly.

Mock Reuben sandwiches go together the same way. Bologna, salami, and canned meats play a kingly role in the sandwich substitute game. They replace corned beef flavorfully, and go along well with kraut and cheese.

The day of artificial food has dawned. What we have now is meat that isn't, milk that could be, and blueberries that really are soybeans. That's progress. With shortages and famine in the world, modern technology gropes for efficiency in food production. Meat analogs (taste-alikes) and vegetable proteins are one part of the solution.

The photos show a synthetic breakfast. The menu calls for scrambled eggs and ham, with blueberry muffins.

The "ham" comes from soybeans processed to give you the texture, color, flavor, and aroma of ham. This meat analog comes in a foil pouch. It's dry, like cereal. You combine it with boiling water to bring out the moist, meat-like properties. Once combined with water, the analog takes on the characteristics of meat. It even spoils like meat. Cost, compared to real meat, is less than half.

The "eggs" are a combination of vegetable substances. The blueberries in the muffins are artificial. You get one healthful bonus with these textured vegetable proteins: little or no cholesterol. The meal is delicious, inexpensive, and easy to prepare.

However, analogs don't solve the problem of sufficient food. Some hope lies in harvesting crops from the sea. But researchers at the Hawaii Institute of Marine Biology doubt this, unless methods of sea farming are developed. Simple fishing cannot produce enough food to feed masses of people. Sea creatures are difficult but not impossible to breed; herds are slow to develop.

Aquatic animal husbandry, sea farming, and protein analogs don't solve food shortages. They may help. Most of the coping is up to us, on an everyday basis.

Chapter 7

Your Share of Scarce Materials

More than 3,000 basic types of products are derived from petrochemicals—one end-product of crude oil. This includes detergents of all kinds, cosmetics, medicines, plastic containers for medicine and for industry, trash bags, vinyl products, toys, dishes, handbags, shoes, crash helmets, cases for transistor radios, television sets, clocks, Band-Aids, baby oil, petroleum jelly, paint, aerosol sprays, agricultural products, newsprint ink. . . . The list goes on and on. If considering all this gives you a headache, count to ten before you take an aspirin—they too come from a petrochemical!

Okay, so what can you do? Most of these products, you can use only once. You can't recycle a Band-Aid or return aerosol propellant to the can. Very few products drawn from petrochemicals can be recycled. Our only recourse is to limit our use of the perishables, and put the rest to work as repeatedly as possible.

The shortage problem could get worse and stay serious for quite a while. Everything and anything you do individually and collectively with friends is bound to help. Through cooperation, maybe everyone can have his/her fair share.

Several products that are in short supply can be recycled. The biggest problem is getting everyone to pitch in. In communities across the country, collection points or bins have been set up. Youth organizations mount drives to end litter. They pick up trash and then sell the reusable treasures to industries. Neighborhood organizations, even local governments, sometimes finance collection programs; trucks make weekly or monthly scheduled collection trips to an area to pick up newspapers, bottles, and cans. Yet, for all that, those are still isolated cases. We need for *everyone* to get involved.

Most families buy a newspaper at least once a week just to get the tv listings and comics. Those Sunday papers are big, and they bulk up quickly. Why throw away even the once-a-week volume of paper when it could be dropped off at a collection point? If you read only Sunday news, one trip every 6–8 weeks would not cause a great accumulation of litter in the house, nor would it tax your personal schedule that heavily. It certainly would help alleviate any paper shortage.

You don't have a neighborhood organization collecting recycle materials? Why don't *you*? Cards posted at the corner supermarket would apprise your neighbors. Your friends should help—even if they only agree to let you take their trash off their hands. Make it a club project. Money isn't the object. It's maintaining a high-quality living standard for everyone (even if it means sharing trash).

Introduce a similar plan for cans, whether tin or aluminum. The way soft-drink cans are tossed around, you wouldn't know that so much cost and effort goes into mining, refining, and forming them. They constitute another resource we are exhausting.

Like papers, cans for recycling should be clean. To conserve space both at home while you're collecting the cans and in the collection bin, do this: After the contents has been used, rinse the can. Remove the labels. Cut out the remaining end. Smash the can flat by placing it on the floor and stepping on it with your shoes. Slip the lids inside. You may wish to step on the can again after the lids are inserted, just to make sure they don't fall out easily in handling.

Natural resources are not inexhaustible. Some took nature millions of years to produce. Too many of us take them entirely for granted. America is still beautiful, where you can find her under the litter. Active recycling may help us rediscover our land.

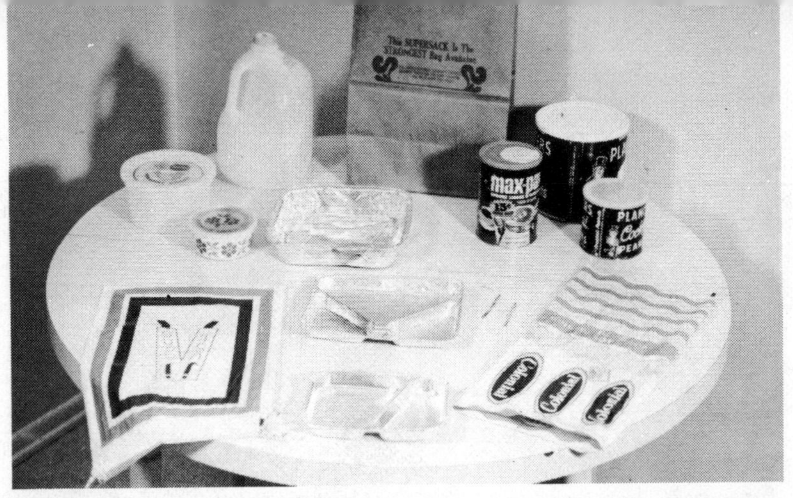

Industry has provided us with mountains of attractive, convenient packaging. We take it for granted. Just about every time you go grocery-shopping, you acquire containers of all sorts—cannisters with freshness seals, see-through plastic bottles, aluminum trays so you can cook without a mess of dishes afterward, paper and plastic bags, and so on.

All this convenience may be headed for extinction. One shortage leads to another. Paper, metal, and petroleum scarcities lead to limits on packaging. Some supermarket managers already request that their patrons save brown paper bags. Some stores can't always get enough bags. (How can they sell their merchandise if people don't have a way to tote it home?)

Plastic containers and metal cannisters with lids make great storage holders. They may not be available much longer, so conserve them now. Why buy refrigerator bowls when you get them free? If shortages continue as long as some predictions say, you'll be glad you saved whatever packaging you could. Utilizing what packaging you are given free could save you dollars.

Here are some can-saver suggestions. Don't like the advertising on coffee and peanut cans? Paint them. Put on decals. Help the kids hand-paint designs for them. If you're not handy with paints, cover the cans with self-adhesive decorative paper. Don't paint the interiors, as the paint might be toxic. Can interiors are specially treated anyway; they won't deteriorate.

Another tip: Save the Christmas-candy and cookie cans for stamp collections, thread and sewing needs, miscellaneous cosmetics, or first-aid supplies. Label the cans for quick identification. Band-Aid boxes, the metal kind, are great for storing tiny items; the size and shape let the boxes fit almost anywhere. Use cookie cans or coffee cans for first-aid kits at home, at the office, or in your automobile and recreational vehicles.

Glass jars and gracefully shaped coffee jars make excellent see-through containers for anything you can push through the mouth of the jar. (The graceful jars are more fun.) Small jars take paper clips and safety pins, or maybe unidentified or extra keys. If you gather things like keys and unidentified buttons in one spot, you'll always know where to put them when you do find them and where to look first when you discover what they're missing from. Nuts, bolts, and screws, which came off something but who knows what, fit this group. Postage stamps store nicely in a glass jar.

Keep glass jars on a shelf with a nautical edge—one that prevents jars sliding off easily. You can glue a thin strip of moulding on the front edge of the shelf.

Everyone needs a place to keep valuable papers or documents. Maybe you don't need a wall safe, but you should have something to prevent loose insurance policies, birth certificates, awards, marriage licenses, and service commendations from being misplaced. Use the cardboard cores from paper towel rolls! Curl the document into a scroll and slide it into the core. Cover the core (which is now a "document case") with kraft or decorative paper. Label each tube individually, and stack them all in the same place. Should a fire occur, they're easy to find and to transport.

You'll find you can use much of what is considered junk. All it usually takes are time and thought. The materials, you find free. The initial effort you spend converting expendables into useful stuff may save you much more time and money in the long run.

We have been thinking about materials for recycling as basically small items. Now think big. Used-furniture stores recycle "junk" at a profit. Beautiful objects, which could be repaired if someone would take the time, are often discarded without much thought.

It's tough to dispose of a mattress in some places. Public auction houses must certify that the mattresses they sell have been thoroughly sanitized and are safe for resale.

If the mattress has been yours, consider repairing it. Holes like the one pictured are okay to repair, if the springs are good. Use foam rubber to replace whatever padding was destroyed. Patch the hole in the mattress casing with ticking or heavy cloth. Hide the patchwork with a mattress cover.

Dressers and chests are easy to repair and refinish. Ask how to do it successfully at an unfinished furniture store or a home center offering paint products.

You can do most minor work needed to square or repair drawers. Nails, plus glue, are the tools. If drawer pulls are missing, buy some attractive new ones. If the furniture has scratches, paint it a light color and antique it. Scratches add character and your piece of "junk" will hold new appeal.

Recycle clothes. If you and yours can no longer wear them, give them to someone who can. Tattered clothes make dust cloths and wipes. Companies that make paper need rags. Don't burn old clothes. They're in demand.

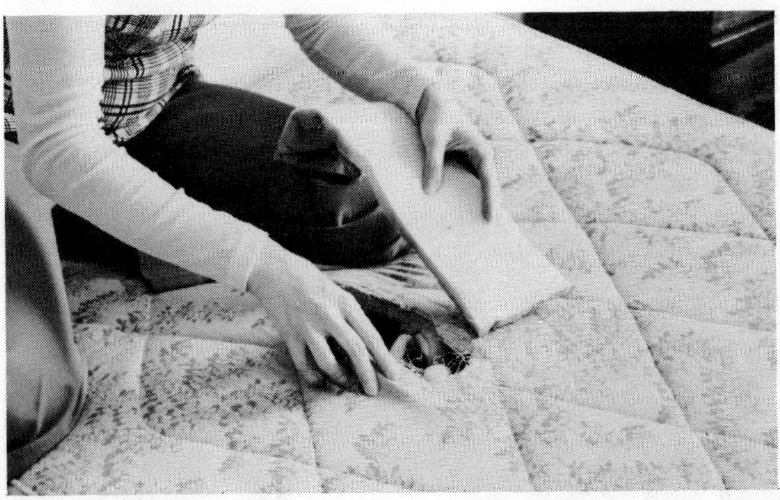

Moving to even bigger things, consider houses. Some fantastic recycling has been undertaken successfully with homes ready for condemnation. Today they're works of art.

Building materials are scarce and costly, things like lumber, electrical conduit, paint, toilet seats, and multitudes of small hardware items. It makes more sense to remodel your present house than to build a new one.

Needs should be considered before desires. If the house you have can be enlarged to accommodate your growing family, you may find that less traumatic than uprooting them. If you have enough space but are looking for more luxuries, perhaps they could be added where you are.

If the kitchen is too small, a specialist in kitchen design could plan you a new one with everything you want (and probably some exciting things you haven't heard of yet). Remodeling a kitchen is generally less expensive than buying a new house.

Home remodeling is a large industry. Check other jobs for quality or workmanship before you sign any contract. Most home improvement companies are honest, but since you will be required to sign a contract, don't take chances. Phone the Better Business Bureau *first.* You may also find that suppliers can place liens on *your* home until the contractor pays his bills for materials used. Play it safe.

If it's just the neighborhood you want to leave, perhaps an older house in a more desirable area will save you money and peace of mind. It's worth looking into, anyway.

We depend on imported raw materials more than is healthy for our economy and our continued well-being. To become more self-sufficient turns out to be a very worthwhile national goal. One way is by making the most complete use of the products we make from these imported materials—till they're absolutely worn out. Another is by finding ways to "run them through again," and get one more use out of them. And, when they're absolutely useless in their manufactured form, maybe we can find ways to reincarnate them as something else.

Several utilities are working on ways to produce *biogas*—a type of fuel extracted from waste. And some companies are treating and compacting trash and burning it as fuel to generate electricity. We need more ideas like these.

A final way to avoid or relieve materials shortages: simply cut down on the stuff we buy and use. Well considered, that might not be as drastic a sacrifice as you think.

Chapter 8

Taking Vacations Despite the Cutbacks

Remember when half the fun was in getting there? You could take the good old RV out for a Sunday picnic, rollicking and frolicking all day. Where have those days gone?

They're still here; they just cost more. Vacation planning has become an art belonging not solely to the travel agent. Despite the shortage of gasoline and other fuels, people still want the most from their free time. The four-day work week will bring even more free time. Can we spend it without consuming additional energy? We can, but the key is persistent investigation. Pleasure boating, private flying, and RVing—everything involving gasoline—take planning.

We have been so mobile in the past, we usually overlooked the attractions in our own backyards. Many fine recreational areas exist near or in large metropolitan areas.

Consider no-gasoline vacations. Why not try backpacking or camping? It might be a break from routine that you would enjoy.

Look to mass transportation. Most bargains have vanished. But popular routes still are served. You can go most places you want to if you plan far enough in advance. The railway system offers perhaps the most leisurely way to see our country.

The energy crisis may very well hand many Americans more leisure days than they'll know what to do with. Even if it doesn't happen directly, the influence will still be felt. Some business researchers report that a four-day work week is likely because it increases individual and corporate productivity. A few companies are now even considering a three-day work week of 12 hours a day. That would save a lot of fuel and energy.

What can you do with that much free time? Free time is like money; the more you have, the more you spend. The secret is to spend your free time constructively and conservingly.

These boys have the right idea. Fishing requires very little of anybody's energy. Getting close to the basics of nature can really give you perspective.

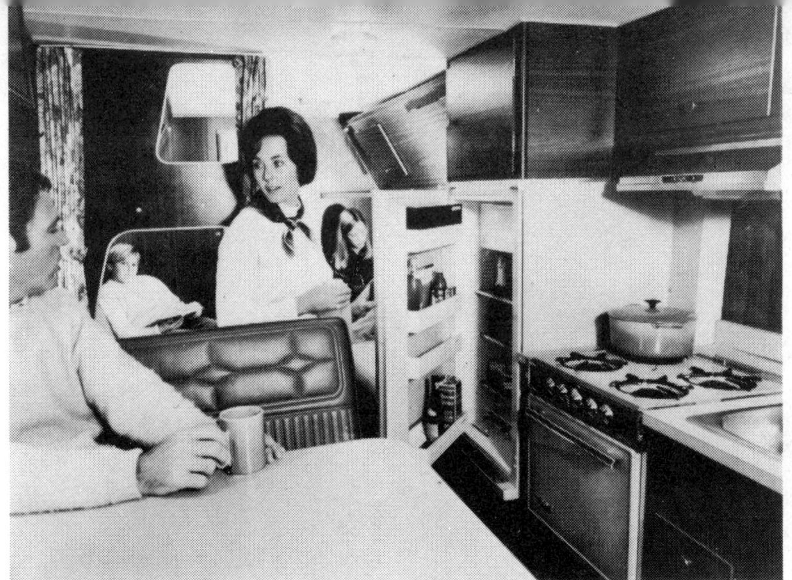

For millions throughout the country, family vacation is synonymous with recreational vehicle (RV). If you own one, you may be among the many who are worried about their investment. How can you enjoy an RV with gasoline so scarce? Leave it on blocks? What happens in case of gasoline rationing?

Large recreational vehicles get rather poor gasoline mileage. Yet, when you spend a weekend in your RV, your family uses less total energy—electricity, heating fuel, etc.—than if you stayed home. Of course, your house can't be "turned off." But, if you set the heat low, it consumes far less energy than if you're there.

The major reason "RVcationing" saves energy has to do with RV lifestyle. You deal in conservation regularly. Your water supply is limited, your electricity comes from batteries in many situations, heating and cooking depend on one or two tanks of lp gas. You have comfort, but you learn to stretch it. Anyone who has not owned or camped in an RV could learn a lot about conservation.

If RVing is your way of life, you probably belong to a car-pool already, to save gasoline money during the week. You could apply the same idea to your RV. You may have friends who would like to join you for a weekend, and share the mounting fuel costs. Or, form a pool with other RV owners. This weekend several of you go in their rig; the next trip, everyone goes in yours. To expand sleeping accommodations, take a tent and sleeping bags or cots.

If gasoline becomes limited enough for rationing, several licensed drivers can share coupons. Two families with four licensed drivers should have no problem working in an occasional weekend RV trip, even with rationing.

Plan your travel days. Fill up your tank before 6 p.m. Any later, you stand a chance of not finding service stations other than truck stops open. Always fill up late Saturday, if you expect to be on the road Sunday.

Know exactly where you're going—no more pot-luck trips. Plan 20-gallon weekends. Pick your campsite so you use only 10 gallons of gasoline to get there and 10 to come back.

Camper parks with all kinds of conveniences and organized recreational areas are readily available. Finding state and national parks within reach of your home shouldn't be too much of a problem.

To conserve even more energy, RVcation in wintertime rather than summer. Head for warmer climates; you'll cut heating costs both at home and away. Maybe a southerly vacation for the family at Christmas would make them happier than individual gifts. Wait until winter to visit relatives in warm climates.

The energy shortage may turn beneficial if we meet the challenges it offers. Figure out how to do the things you want to, in spite of shortages.

Don't be afraid to use your RV. If everyone who owns an RV would take it out for weekends more often, they would save more energy and enjoy doing it.

Most people think of vacations in terms of faraway places and exciting activities. However, you can probably find a variety of fascinating places to visit near home. A bit of checking might turn up points of interest you never even realized were there. The country is full of historic monuments, museums, art galleries, observatories, aquariums, zoos, any number of interesting things to see and do.

Look over maps of your city and state. If the ones you have don't indicate spots of interest, order those that do. Some oil companies put out fantastic recreational maps. Also query your state Department of Tourism. If your state doesn't have a tourist bureau, try the Department of Commerce.

Gather the whole family together to discuss what's available nearby. Children reap special benefits from a broader view of their home territory, because social studies in school often emphasize events and places of local and state significance.

It's surprising how many people don't know other sections of their own city. One efficient way to acquaint yourself with any area you don't know is by a guided tour. See everything you can quickly; then go back on your own and spend time at places that interest you most. Take your family on a guided tour of your own city. If there are none, take them on one you've planned. See the entire city—its beautiful monuments, its architecture, its slums, its museums, its parks. The trip could be enlightening.

Get to know your own city and state first, especially now that gasoline is so scarce. By the time you have soaked up all there is to know about home, the energy problem may be over.

Your state parks and recreational areas can stretch your weekends and save driving. Too many weekend "vacations" are spent on the road. You finally arrive Saturday evening, tired and ready to go to bed. If you sleep late Sunday morning, your weekend is gone before you have time to enjoy it. Don't go so far afield. State parks and resort areas are interesting and relaxing. They exist for your convenience and pleasure.

The maps you picked up from your service station or state tourist bureau usually have a legend that identifies facilities each area offers—camping, bicycling, swimming, fishing, boating, horseback riding, type of room accommodations.

Make reservations for rooms or camper space several weeks in advance. Estimate your time of arrival. Start as early on Friday as you can. Plan to get there before nightfall; that leaves you almost two whole days to relax.

Diversify your activities during the two days. On Sunday, attend an outdoor church service; you may find it rewarding and renewing.

After a little practice, you'll find your weekend trips bring you more relaxation than ten-day vacations on which you drive several hundred miles. Those faraway resort areas seldom offer any more chances for peace and quiet than you can find right around your home state.

Another way to make a weekend seem like a ten-day vacation is by flying. Pick a spot to which you can fly in a couple hours. Leave on Friday. Plan two days of specific activities or rest. Then go.

A weekend trip can seem like a full vacation on longer flights. Choose someplace vastly different from where you live. From the midwest, try a Caribbean weekend. The change is like walking through the looking glass! Cram all the sightseeing, shopping, sunning, and funning you can into the two days. Rest on the return flight. A fun-packed two days can restore you like weeks might. If you have an extra vacation day coming, save it for a trip like this. You'll love that extra day.

For any short vacation, fly. Many special fares have been dropped from air travel schedules, but you can still find economical ways to fly. The best bargains now are the fares on the night coach flights. These rates apply to flying after 10 p.m. and are approximately 20% cheaper than day coach fares. Adults fly for full fare, children 2–12 years old go for two-thirds fare. When you make reservations, ask for the night coach rates. And don't wait until the last minute; these flights book up rapidly.

If you really want to spread out a short vacation, try the fly-drive idea. For example: Fly to Miami or Jacksonville and rent a car or camper. The airline you fly with can make these arrangements for you. Cover all the sights you want to see, and catch a flight home from Tampa. That avoids several days driving time, which alone will make your vacation seem longer than it was. You'll also save a lot of gasoline expense.

No matter where you want to vacation, the fly-drive plan has advantages. Possibilities exist worldwide.

Partly because of the energy crisis, the iron horse has been brought back from pasture. Amtrak, the new government-subsidized railway system, has been busy upgrading and refurbishing passenger train service.

One of the rewards of a train vacation comes when you pass through the most scenic regions of our country. The educational benefit, and just plain visual pleasure, cannot be overstressed for children or adults.

Train travel on Amtrak is either by coach or pullman. Amtrak has a family plan for coach accommodations. Dad pays full fare, Mom and children 13–21 go for two-thirds fare, children 6–12 go for one-third fare, and children under 5 go free (except for a minimal seat charge).

Coach service on these trains does not include sleeping berths. Instead, you sit in reclining seats. This may sound a bit uncomfortable. But considering the stress of driving, this probably leaves you more rested when you reach your destination.

Rail trips take about as much time as driving would. Trains today average 50 mph including stops. A trip from coast to coast takes 3½ days.

Meals on an Amtrak train are reasonably priced. Breakfast ranges from 90¢ to $2.50. Lunch runs from $2.25 to $3.50. Dinners are seldom over $5.00. You can also order child's portions. Fares aren't too bad, although prices could go up anytime. A sample coach fare: Indianapolis to San Francisco, one way, for under $100. The train would leave Indianapolis at 8:45 a.m. on, say, Monday, and arrive in San Francisco on Wednesday at 3:05 p.m. This particular route takes you through what are possibly the most spectacular regions in America. At $100, it's cheaper than most people can drive their automobile. The train at 50 mph arrives just as fast, and you aren't exhausted by the long auto trip.

If you're going to Florida, consider Amtrak's Auto-Train. On it, you take your automobile with you. You save the gasoline it takes to get there. However, these special trains are not widely used; phone the Amtrak office in your town for details, routes, and schedules.

Perhaps you've never considered a bus for vacationing. But, since transit vehicles are expected to receive ample fuel, you might be wise to use their fuel instead of your own.

Major bus lines have a variety of vacation programs. They include tour-of-the-month, special-price fares from time to time, domestic and world tours. Tour prices include bus transportation, hotel lodging, certain meals, and tips for baggage-handling.

One of the best buys in bus travel right now is the $149 ticket that is good for 30 days and takes you almost anywhere you want to go in the U.S. and to parts of Canada. (Price subject to change anytime.) Quite conceivably, you could see the entire country in a month. It would be once-over-lightly, but you could visit a lot of places. If you feel two months is more like it, you can get an extension pass for $50 more. These 30-day tickets are not tours. You decide where you want to go, and where and how long you want to stay. You make your hotel arrangements and buy your own meals. However, some of these passes entitle you to cut rates at some hotels and restaurants.

If you want to discover the U.S.A., and you have a limited budget but lots of time, this is one way to manage it.

Dress casually. Don't weigh yourself down with too much luggage; carry one bag with you. Take a friend along.

One last vacation suggestion. Try a vacation at home. You might be surprised what a comfortable time you could have. Turn off the telephone bell; let the newspapers collect on the front porch; don't answer the doorbell.

Plan this kind of vacation like you would any other, months in advance. If you want to plant the garden, take your home-vacation in late April. You can plan vacation time from the moment the seed catalogs arrive. You may want to test the soil early, so you can stock up on the additives and plant foods you'll need. Do you have enough stakes for the tomato plants? And where did you put the tools last year?

Perhaps all your other leisure time has been consumed in a multitude of activities. Maybe just reading, painting, sewing, or woodworking are more to your liking. Plan it a few months in advance. Get together the tools, materials, books, or what-have-you. You don't want to waste vacation with shopping.

If you think your youngsters would get bored on a stay-at-home vacation, let each one plan what he/she would like to do individually. (One may decide to read your books.)

Whatever you decide plan ahead. Then enjoy, enjoy, enjoy!

Most important of all: Don't let the gasoline/energy/food/etc. shortages scare you out of taking a vacation as often as you want to and can afford. Frequent rest and recreation leave you in the mood to cope with the "crisis" the rest of the year.

A vacation? You owe it to yourself.